# The AMATEUR NATURALIST

## and Exotic Petkeeper

### Issue Nine

# Explore the world before it's too late

Edited by Max Blake
Typeset by Jonathan Downes
Cover and Internal Layout by Jon Downes and spider-oh for CFZ Communications
Using Microsoft Word 2000, Microsoft , Publisher 2000, Adobe Photoshop.

First edition published 2010 by CFZ Publications

**CFZ PRESS**
**Myrtle Cottage**
**Woolfardisworthy**
**Bideford**
**North Devon**
**EX39 5QR**

THE CENTRE FOR FORTEAN ZOOLOGY
www.cfz.org.uk

**ISBN: 978-1-905723-63-8**

# Plea In Mitigation

Hello again, and welcome to another issue of *The Amateur Naturalist*. And the digital edition at least made our big deadline set last issue, that is, to get this issue, issue 9, out before the AES show.

I am happy to announce that next issue, whenever it comes out, will be the fourth issue of *The Amateur Naturalist*, but it will be the tenth issue if one includes *Exotic Pets*, so I think that this calls for something special. Indeed, we are going to be keeping things close to home and make it a European special edition. That means that everything you read within the magazine (apart from reviews, but we will try) will be about European topics. European news, European exotic pets, European conservation and European natural history will all be featured. We do try to include European topics in every issue of *The Amateur Naturalist*, this issue we have pine hawkmoths, Irish wolves, the natural history of your garden and the changing herpetofauna of Denmark amongst other things because as "doing the green thing" continues to be a major buzz word, the editorial staff of this magazine would like to encourage more people to do things close to home (seeing as most of our readers are from Europe) to reduce such things as carbon footprints and air miles.

Speaking of the staff, it is with sadness that you won't find any articles or columns by Graham Smith, Trevor Smith or Janice Holt in this issue. Why? Because Graham and Trevor are currently involved in a major project which is proving to be extremely time consuming. I don't think we can announce what exactly it is they are doing, but hopefully, if everything goes well, when they release their findings you will understand perfectly why they were unable to contribute to this issue. And trust me, when they release what they have got so far, every natural historian in Britain will sit up and take notice. So, I wish the both of them luck with their escapades and

In the last issue I relinquished editorial control to Max Blake (currently just about to start his second year of studying zoology at Bristol University) and made him Guest Editor.

He did so well that I have expunged the word 'Guest' from his job description, and he is now the editor.

I have immense faith in him. In the two and a half years since he became involved with the CFZ he has become increasingly important both to my wife Corinna and myself, and to the CFZ in general, and I cannot imagine life without him. He has a great future career ahead of him, and I confidently predict that he is going to be a very bright star indeed in years to come.

Watch this space.

Jon Downes

# Close to the Edge

I hope their project brings them as much success as I think it will.

With this issue of TAN, we are aiming to start really bringing the magazine to a wider audience. I'm not talking about running thirty second adverts in the middle of *Coronation Street*, none of us would want to encourage that sort of tripe, but we will be aiming to send copies of this very magazine off to school libraries up and down the country to help bring an education in Natural History to those who can make the most difference in the future: children.

Natural History has never really been taught as part of the school curriculum but would have been taught, back in the day, by parents taking children on walks in the countryside and the like. Over time, this has been phased out in favour of other more gaudy attractions, so, far from suggest that every parent should be forced to take their children out on walks every Sunday without fail, we want to see children *asking* to be taken on walks and such like, which is hopefully where this issue comes in. It is not a guide to everything you can find in your garden, there are plenty of books on that topic, but it is a series of mini guides to different aspects of Natural History, including something which is very easy to do in the modern age: keeping and breeding your own exotic animals and watching their wonderful behaviour.

One of the biggest successes that we have had this year is that of orchid conservation. Every year, the early purple orchids (*Orchis mascula*) situated by the A39, just outside a small village called Fairy Cross, are cut down before they can fully release their seeds. These, like many other orchids, are scarce in the UK, and so we decided that this year, it was time to do something about it.

After letters being sent from the CFZ to the council, back again, forward again, and back again, the council finally officially put the small verge containing the flowers in the hands of the CFZ to look after.

So for the first time in years, the early purple orchids at Fairy Cross have been allowed to fully shed their seeds.

We had ten flower spikes come up this year, and so naturally, we will report in this magazine during the next season (hopefully sometime in April or May) just how they get on.

Here - by the way - we would like to

thank Paul Haresnape for his generous donation of funds to pay for a petrol strimmer and the protective clothing which is now a legal requirement for those working on a road verge. He made the donation in the name of his late mother, a fellow of the Royal Horticultural Society and noted painter of orchids.

Recently I have had a number of people, more than usual anyway, ask me "Why is the magazine not in colour? It means we can't see what the animals look like in the flesh!" to which I say "Did you actually bother to read the contents page where it says "full colour digital edition"?" Well, admittedly, not many people would bother reading what looks like a very dull legal piece and so I can't really blame anyone. But, hopefully more people read this editorial and so I would like to remind everyone that if they have bought a hard copy of the magazine, they are entitled to download the digital, full colour version for nothing at all.

Just email Jon (email address on the right) with proof of purchase and he will send you a download link. So remember, if you want to contribute something to the magazine, but you are worried about the beautiful colours in your photos getting monochromed, don't worry!

The digital edition will still preserve their beauty. That is, unless we get a huge amount of demand for a full colour version of the book, which would retail for around £10…

Max Blake (Editor)

**Max Blake**
(Editor)
max@cfz.org.uk

**Jonathan Downes**
(Editor)
jon@eclipse.co.uk

**Graham Smith**
(Contributing Editor)
graham.smith19@virgin.net

**Corinna Downes**
(Sub Editor)
corinna@cfz.org.uk

**Richard Freeman**
(Herp Editor)
richard@cfz.org.uk

**Oll Lewis**
(Staff Writer)
Oll.lewis@cfz.org.uk

**The Amateur Naturalist**
Centre for Fortean Zoology [CFZ],
Myrtle Cottage,
Woolfardisworthy,
Bideford,
North Devon
EX39 5QR
Telephone: +44 (0) 1237 431413

Payment by cheque drawn on a UK bank account (payable to CFZ Trust) or by Paypal using our account: jon@eclipse.co.uk.

**FOUR ISSUE SUBSCRIPTION**
(including p&p)
£18.00 UK
£ 20 (€ 32) Europe
£25 ($US40) US/Canada
£30 ($Aus60) OZ/NZ
£30 Rest of World

# Contents

## CONTRIBUTORS

Jon Downes, Corinna Downes, Scottie Westfall,
Richard Muirhead, Lars Thomas, Carl Portman,
Richard Freeman, Nick Wadham, Mark Pajak,
David Marshall,  Oll Lewis, Lucy Henson, and
Max Blake

## DISCLAIMER

## FULL COLOUR DIGITAL EDITION

This magazine is available in two formats. The
perfect bound paperback format costing £4.99/
$US8.99 and the digital format costing £2/$US4.
If you have purchased the hard copy format you
are entitled to have a free digital copy.

This is a service of the magazine and its publishers who realise that although the hard copy is
more durable and looks well on your bookshelf,
the contents are in black and white, so email
info@cfz.org.uk with proof of purchase, and
download instructions will be sent forthwith.

# BugFestSW
## BUGFEST 6

The next Bugfest will be held in February 2011. After two years of holding a bi-annual event we have decided to concentrate our efforts on a yearly show, bringing the best in creepy crawly fun to you. All details will be posted on this web site nearer the time, so keep checking us out!

---

## What is Bugfest?

Bugfest is an entomological (creepy crawly) exhibition held at Bucklers Mead Sports Centre, Yeovil, where traders from across the country exhibit, sell and offer advice on invertebrates from crabs to stick insects. There are also lots of other stalls including those selling books, crafts, insect housing and children's craft stalls. There is also a children's 'spot the bug' quiz with small prizes to be won! Smaller workshops are regularly held during school holidays which are primarily aimed at children who can come and meet Nick and his collection of bugs, ask questions and handle insects. Nick also visits schools and groups to talk about his ever growing collection of creatures.

## Who runs Bugfest?

Bugfest consists of a husband and wife team, Nick and Kara Wadham, ably assisted by family, friends and volunteers, who decided to bring the creepy crawly world to Somerset. The popularity of the first event was a complete surprise as 1000 visitors attended. Now Bugfest is going from strength to strength, drawing notoriety and attention to the South West!

# RSPCA MAKE JUMBO FOOLS OF THEMSELVES

The Royal Society for the Prevention of Cruelty to Animals (RSPCA) has scored a spectacular own goal recently by calling for the import of Elephants to UK zoos to be stopped. On the face of it this might seem a reasonable request, however, it shows that the RSPCA, or at least the member of the society responsible for this decision, know close to nothing about genetics, conservation, zoos and simple biology that one might learn even at a pre-GCSE level. There are only 70 elephants in 13 zoos in the UK, so if importation of new elephants were to cease, then this comparatively small population would become genetically isolated, resulting in a genetic bottle-neck through enforced inbreeding. Genetic bottlenecking causes a population to be less adaptable to change, have less resistance to diseases and for genetic problems to be amplified within a population.

Bottlenecks occur within nature as the result of enforced isolation of a population or in the aftermath of natural disasters and other cataclysmic events. When a bottleneck occurs it is essentially the wiping out of a large amount of the potential genetic variation within the species. To use an analogy, it would be the equivalent of an author taking a dictionary, ripping out all entries that begin with a letter in the first quarter of the alphabet and then writing a novel, using only the words he has left in the dictionary. It may be technically possible, it may even turn out to be a good book, but it would be certain to have problems. When the available genes drop below a certain level the population may no longer be genetically viable in the wild, which is why conservation biologists keep stud books to ensure that as wide a genetic diversity as possible is maintained when breeding endangered species in captivity.

The RSPCA released its statement in response to a report to DEFRA by the Zoos Forum Body into Elephant Husbandry (http://www.defra.gov.uk/wildlife-pets/zoos/documents/elephant-forum-1007.pdf), a body that the RSPCA were invited to be a part of but declined.

Dr Ros Clubb from the RSPCA said the charity was disappointed the report did not recommend an outright ban on importing elephants to UK zoos. Clubb said: "The RSPCA believes that until solutions to the extensive and serious welfare problems can be found, we should not be introducing more elephants.

"Elephants are, without question, suffering in zoos.

"Adding yet more elephants to an ailing population simply masks the problems and if drastic improvements to these problems cannot be found, the RSPCA believes zoos should phase out elephant-keeping."

The RSPCA seem to have ignored the fact that one of the purposes of the report was to ensure standard practices to improve elephant husbandry would be implemented by law, which would effectively put an end to any elephant suffering.

The RSPCA seems also to be unaware of the need for captive breeding of elephants for use in future reintroduction programs. Whereas African elephant populations are growing thanks to the efforts of various national parks and zoos, the ban on ivory sale and more serious control of poachers, the numbers of Indian elephants are in free-fall.

Richard Field, the director of Whipsnade Animal Park, said there were a number of reasons why elephants and other endangered species needed to be kept in zoos. "It may be that they are being bred for future reintroduction, it may be that their populations in the wild are going through dramatic declines - as is the Asian elephant - and we need insurance populations."

He said the knowledge gained about elephant behaviour was then used to help help preserve the animals in the wild. "We need the RSPCA to help and work with us, not refuse to join with us."

Georgina Groves, who is overseeing the auditing of all the UK and Irish elephant collections, said: "BIAZA are surprised and disappointed by the RSPCA 's disregard for the Zoos Forum's expert review carried out on the Bristol University research of elephant husbandry in UK zoos, and their consequent call to stop the introduction of new elephants into the UK."

The RSPCA do a lot of good work in helping to prevent animal cruelty but they should be careful to get their facts straight, and research fairly basic science before they let their hearts rule their heads and suggest measures which would have a detrimental effect on animal populations in the future and hamper conservation attempts. They are not PETA (who want a ban on all zoos) and should not act as if they are. The

RSPCA should attempt to work within the system, as they were invited to here, rather than acting indignant and trying to rubbish a good solution to a possible animal welfare solution that should be enough to satisfy all parties.

## THE PLIGHT OF THE ASIAN ELEPHANT

Whereas efforts to save African elephants are paying off, for some subspecies of the Asian elephant (*Elephas maximus*) it is quite a different story. The Indian elephant is a jungle dwelling subspecies, in contrast to the African elephant which lives primarily in fairly open savannah; this makes it harder to keep track of Indian elephant numbers and to identify when and where poaching may be occurring. Poaching is not the only threat that Indian elephants face; threats come from the fast increasing population of the Indian subcontinent and Asia, coupled with an increase in India's fortunes. Not only is more land needed to house and feed the growing population but often if a person has made money they will want a larger home and more land.

Often this land will come from the elephant's habitats leading to habitat destruction. Elephants, that used to be seen as useful animals are increasingly being seen as pests that will destroy a farmer's crops. There are thought to be as few as 20,000 Indian elephants still in the wild, which - because of their wide distribution - has resulted in a fragmented population.

Without intervention by conservationists it is possible that these populations could dwindle and become more genetically isolated resulting in the future extinction of the subspecies. Charities like the World Wildlife Fund are helping by setting up reserves and giving police more equipment and funding to help catch poachers but zoos can help as well by ensuring a greater genetic diversity is maintained in smaller populations.

## GOLDIE LOOKING FISH

Back in July, apparently, some photographs appeared on various specialist fishing websites and forums showing a giant "goldfish" that had been caught in the south of France. At the time of going to press, the *Daily Mail* printed the story and other websites were quick to link to it and to pass it on, and it eventually reached Ruby Lang, in Australia, who forwarded it to me (Ed.). The picture attached shows a very large (30lb) golden cyprinid being held up by the angler that caught it, Raphael Biagini. Fellow anglers in the area had told Biagini that they had spent the last six years trying to catch the fish, but without success. After eventually getting a bite from the fish, Biagini fought for ten minutes to bring it in before weighing it. Anyway, enough generic reporting…

Golden sports of fish rarely survive in the wild, so it is impressive that this one (the largest golden carp caught in the wild) has lived to such a great size. There are comments on various articles denouncing the photograph as a fake; however the fish and its proportions etc look normal to me. The fish is being held away from Biagini in typical angler fashion to exaggerate its size, which is nothing unusual, and you can see this slight distortion with the wide angle lens used in the size of Biagini's fingers: those toward the fish's head, and closer to the camera, are larger than the fingers at its tail. According to some comments, this means the photo is doctored. Right...

Apparently, there is also no bend in the part of

the caudal fin under water, which makes the photo a fake. Again, there is a clear bend in the fish's tail which bends the tail slightly upwards toward the surface of the water. In other words, exactly what one would expect. However, his head seems huge compared to his knee, and he is holding the fish with very little strain for something which weighs the same as a three-year old kid...

I would be tempted to say that the 30 pounds quoted was an exaggeration, but the fish is still large; it has the barrel chest that carp get as they approach big sizes. Whether the carp is a totally wild golden sport of a normal carp (*Cyprinus carpio*), or a feral koi I cannot tell, but suggestions would be most welcome.

As an interesting side note, koi carp were developed from wild east Asian carp, *Cyprinus carpio haematopterus*, in around the 1820s after being kept in the wild colour for centuries (possibly before the time of Christ). However, goldfish were first developed from Prussian carp (not Crucian carp as is so commonly suggested) over 1000 years ago and are now labelled as a different species from the Prussian (*Carassius auratus auratus* vs *Carassius gibelio*), though this is likely to change. The theory that goldfish are descended from Prussian carp is fairly young, as theories go, and the taxonomic changes surrounding it have not really taken place. Thus it seems likely that the Prussian carp will be renamed *C. auratus gibelio* because the goldfish was named first (1758 vs 1782), and under the rules of scientific nomenclature, the oldest scientific name for a species is the one that stands.

## FERN-TASTIC

Botanists at London's Kew Gardens have rediscovered a species of fern from Ascension Island and successfully cultivated plants in their laboratories. The fern, *Anogramma ascensionis*, was thought to be extinct since the 1930s until it was rediscovered clinging to a steep mountainside by Kew botanist Phil Lamden and local

conservation officer Stedson Stroud.

"We were down the back of Ascension's Green Mountain, which has very, very steep slopes. You have to be really careful because if you slip you're a goner," Stroud recalled.

"And we came across this beautiful little fern and immediately knew it was the lost *Anogramma* that had been extinct for the last 60 years."

"We were scrambling around, looking to see if there were more, and then we realised, we should really have safety ropes and stuff around us."

Along with their college colleague Olivia Renshaw, the scientists managed to find only four more of the small plants and mounted an effort to protect them through a dry period.

"We had to keep the plants alive - they were on a bare rock face and it was a really dry period, so Olivia and I went down twice a week carrying water and we set up a drip feed." said Stedson.

After protecting the plants through the dry spell the next stage was to take cuttings from the plants back to the Kew laboratories in the hope of cultivating them to increase the number of plants. Cuttings from the plants, containing spores, had to be kept moist and transported to Kew within 24 hours where Dr Viswambharan

Sarasan could bleach the spores to kill any bacteria.

"That is the really risky part," Sarasan said. "If you bleach them for too long, you could kill the spores, but if you don't treat them for long enough, there could be remaining bacteria that will grow in culture and kill them."

Sarasan and his assistant, Katie Baker, have successfully cultivated 60 new plants which botanists hope to eventually restore to Ascension Island. Stroud has also successfully cultivated some *Anogramma* himself in a shade house on Ascension Island.

"Each and every day, you're there, tending and looking, and hoping that something will happen," Stroud said.

"Then one day you see something and - watching the plants grow - you can't ask for anything more."

## OIL GET MY COAT

On the 20th April 2010 the Deep Sea Horizon oil-drilling platform, situated in the Gulf of Mexico, approximately 40 miles off the coast of the USA, was drilling for oil in the deep sea when it exploded, resulting in the death of 11 workers. The explosion was tragic in itself, but what was to come would prove to be even worse.

The explosion and subsequent sinking of the drilling platform meant that there was a gaping hole deep on the seabed, through which oil was shooting out at high pressure into the Gulf of Mexico. The warm tropical sea of the Gulf is home to a very diverse and unique marine ecosystem, not only providing a permanent habitat for many species, but also breeding and feeding grounds for many other species. Estimates of the amount of oil flowing into the Gulf per day varied widely from BP's -- the well's operator - initial estimate of only 1000 barrels a day, to the estimates by the Flow Rate Techni-

cal Group - a team of scientists recruited by the Federal government - of between 53,000 to 62,000 barrels a day. Whichever estimate was the most accurate the result was an oil slick that covered a large swathe of the Gulf, encompassing a surface area of over 10,000 square kilometres.

Eventually, after several failed attempts, BP was successful in capping the last of the oil plumes in mid to late August and the American government issued a statement claiming that most of the oil had been dispersed or evaporated. However what the government's hasty statement failed to take into account was that so called 'dispersed oil' is still present in the water column, but is harder to see or keep track of. Scientists from the University of Georgia estimated that as much as 80% of the oil could still be present and that to claim that dissolved oil is gone is a gross misrepresentation of data.

More than 400 animal species, classified as being at risk or lower, are threatened by the oil spill including already endangered species like green turtles (*Chelonia mydas*), Kemp's Ridley turtle (*Lepidochelys kempii*), the loggerhead sea turtle (*Caretta caretta*), the hawksbill turtle (*Eretmochelys imbricata*) and the leatherback turtle (*Dermochelys coriacea*). Thousands more species will also be effected by the slick and the after effects, including several species that are unknown to science like the two new species of pancake batfish that were discovered just before the oil spill. By the 13th August, 4,768 dead large animals had been collected according to figures reported to the U.S. Fish and Wildlife service.

The dispersal chemicals used during the 'clean up' operation have also caused the oil to be broken down into small enough particles to enter the food chain, through filter feeders which will result in more animal - and possibly human - deaths to come.

The US government and BP's response to the oil spill has been largely focused on getting rid of the visual signs more than the long term ecological impact, hence their eagerness to declare 'most of the oil gone' when it is still present but dispersed in the water. The worry is that with the majority of Americans believing their government's misleading press release, and many media companies seeing this as a good opportunity to draw a line under the Gulf oil spill as a 'completed story', the ecological aftermath that is still to come will not get as much funding as it should due to lack of interest and, therefore, will have to be done on the cheap. That would exacerbate this tragedy even further and add to the death tolls.

## PITCHER PERFECT

Scientists have discovered the smallest adult frog in Afro-Eurasia on the island of Borneo. The frog has been named *Microhyla nepenthicola* after the plant it lives in.

Adult males of the frog only grow to between 10.6 and 12.8 millimetres in size and the tadpoles live in water found in the pitchers of tropical pitcher plants.

Dr Indraneil Das of the Institute of Biodiversity and Environmental Conservation at the University Malaysia Sarawak said the species had originally been incorrectly identified in museums.

"Scientists presumably thought they were juveniles of other species, but it turns out they are adults of this newly-discovered micro species"

The scientists discovered the minute frogs after following their harsh rasping calls alongside a road leading up the slopes of Gunung Serapi Mountain in the Kubah National Park.

## LITTLE GIRL'S LUCKY TIGER ESCAPE

A mother has revealed that she is unlikely to renew her family's annual ticket to Jungle Is-

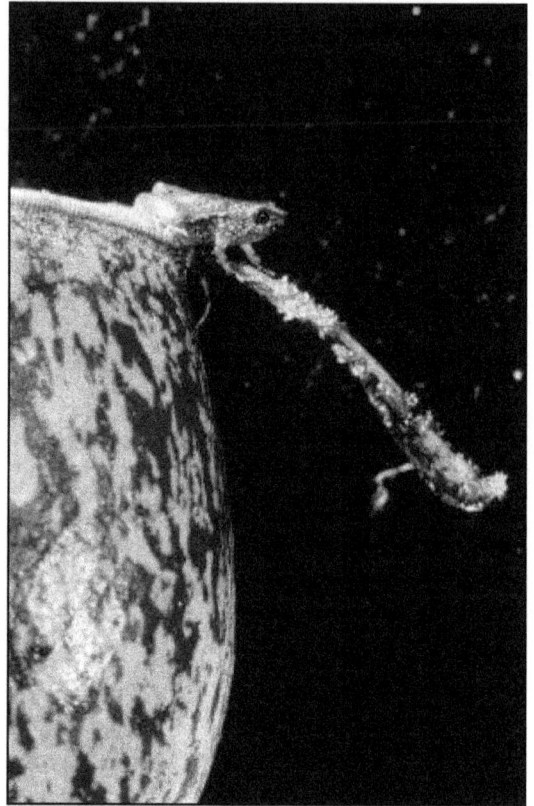

land Zoo in Miami, after her two-year-old daughter came face to face with an escaped tiger. The 230kg Bengal tiger named Mahesh managed to jump over its enclosure's 12 foot tall fence and wandered around the zoo for around 20 minutes before it came face to face with young Dianita Barret.

Dianita's mother, Diana Barrett, said: "I rounded the corner and she's standing there staring at this tiger about 10 to 15ft away from her just standing there not moving, neither two of them were moving."

"Well the tiger was very calm and I didn't want to make it anxious at all so I took a few steps towards her, picked her up, and walked in the other direction."

Diana quickly got her daughter out of harm's way and keepers caught the tiger.

The zoo claims that the tiger escaped after being antagonised by a 'mischievous gibbon' who had also escaped.

This is not the first time recently that a tiger has escaped from an enclosure in an American zoo; on Christmas Day 2007 a tiger escaped from an enclosure in San Francisco Zoo after being taunted by a young man, and after scaling an 18ft tall fence went on to kill it's tormentor and several other people at the zoo. Thankfully, nobody in Miami was hurt but the zoo's apparent lax attitude towards safety will result in harsh penalties, hopefully ensuring that the zoo make sure that all enclosures are secure in the future. After the San Francisco incident made the headlines, it seems like madness that the Miami Jungle Island Zoo did not improve their own tiger enclosure, especially as their fence was 6ft shorter than the one at San Francisco, and easy to escape over by a determined tiger.

## LURKING LOBELIA

A species of lobelia, thought to have been extinct for almost 100 years has been rediscovered. The flowering plant was found on the slopes of Kohala volcano on the Big Island of Hawaii by researchers and conservation workers including Jon Griffin, who described the moment of discovery:

"We were surveying a rare tree snail population when we came across a native lobelia plant that we were unable to identify."

Griffin and the other conservation workers found about 30 other specimens of the plant and sent photographs to University of Wisconsin Oshkosh botanist, Dr Thomas Lammers, who identified the plant as *Clermontia peleana singuliflora*. The species, which was known by the local name of Oha Wai, was last seen on the island in 1909 and last recorded in 1920 on the nearby island of East Maui and has never been recorded on Kohala volcano before.

The plant has greenish white flowers and dark green leaves with red under ribbing. Conservationists are hoping to propagate the species to ensure its survival.

## CHATTANOOGA ZOO GOO

A team of keepers from Chattanooga Zoo, Nashville Zoo and researchers from Michigan State University and Lee University have launched perhaps the most determined effort yet to get hellbender salamanders to breed in captivity.

Building on successful freezing and thawing of salamander sperm that remained viable when the process was completed last year, Chattanooga Zoo's lead ectotherm keeper, David Hedrick, and curator of ectotherms, Rick Jackson, have met with several other experts to collect hellbender sperm from around the Hiwasse River for preservation.

As well as using the frozen sperm packets in future artificial inseminations, the keepers and scientists hope to recreate the natural environment of the hellbenders as accurately as possible to try to facilitate natural breeding between their captive hellbenders and to determine what exact conditions trigger the breeding of the large salamanders.

Hedrick said: "We're coming into crunch time with this species. The next 20, 30 years are going to determine what's left in the wild, if much of anything.

"They like clean, clear, swift-flowing cool water, and they're really a fabulous indicator of water quality."

Jackson commented: "What tends to happen is there are more people out looking at cougars and your larger mammals. When you go to reptiles and amphibians, a lot of times it's just a lack of researchers."

Also present on the sperm packet collecting trip was Michigan State professor of reproductive pathology, Dalen Agnew, who was very optimistic about the conservation project:

"What I love is that it's a great collaborative effort, we've got people from Europe, from Nashville, from Michigan, from Chattanooga all working together for a pretty cool species."

## ARE HELLBENDERS UNIQUE

Hellbenders are the largest known salamanders in the New World. Together with the two Asian giant salamanders they make up the family Cryptobranchidae. But could there be other North American giant salamanders, possibly even larger than the hellbender which reaches a whopping 29in?

In his book *Tom Slick and the search for the Yeti* (1989) Loren Coleman describes some giant salamanders reportedly living in the Trinity Alps of Northern California. One elderly witness told Oilman/Cryptozoologist Tom Slick that in his youth he had seen salamanders "the size of alligators"!

There are also persistent rumours of giant salamanders the size of hellbenders living in remote lakes in British Columbia.

# A PRINCE AMONG MEN

London Zoo recently solved a small mystery; why their two-toed sloth, Prince, was refusing to mate with the sloth they had brought in to partner him. Keepers had high hopes that the two sloths would mate when they introduced Sheila into Prince's rainforest-style enclosure and that the zoo would be able to successfully breed two-toed sloths for the first time in over 100 years.

Sadly, it was not to be and after trying several different tactics to get the sloths to show interest in each other, like laying scent trails and trying to ensure that the sloths moved around the enclosure more to maximise the number of times they would have to interact with each other, nothing seemed to work. In view of this, the keepers investigated possible reasons why the sloths steadfastly refused to mate. As part of this investigation they brought Sheila in for an ultrasound scan, where they found their answer: Sheila was male.

Sheila has now been shipped out to another zoo and replaced with a three-year-old sloth called Marylin - who is definitely female. Keepers are hopeful that the new pairing will prove to be more successful.

*Choloepus* or two toed sloths is a genus of mammals of Central and South America, within the family Megalonychidae. There are only two species of *Choloepus* (meaning "lame foot"):

Linnaeus's Two-toed Sloth (*Choloepus didactylus*) and Hoffmann's Two-toed Sloth (*Choloepus hoffmanni*).

Although similar to the somewhat smaller and generally slower moving three-toed sloths (*Bradypus*), there is not a close relationship between the two genera. Recent phylogenetic analyses support the morphological data from the 1970s and 1980s that the two genera are not closely related and that each adopted their arboreal lifestyles independently.

# CRANES LAY SIEGE TO ENGLAND

A siege of cranes will be seen over the English countryside for the first time in 400 years, as part of a reintroduction project by the Wildfowl and Wetlands Trust (WWT) in Slimbridge.

The 'Great Crane Project' is the culmination of years of work by the WWT, the Royal Society for the Protection of Birds (RSPB) and the Pensthorpe Conservation Trust. Eggs were shipped in from Germany in April and hatched at WWT Slimbridge, where the staff wore crane costumes to ensure that the chicks did not imprint on humans. The birds are currently in a temporary release enclosure at a secret location in Somerset. WWT's head of conservation breeding, Nigel Jarrett, said: "We'll be doing what is known as a soft release - meaning we do it very carefully and slowly so the birds hardly notice it's happening.

"We have to be comfortable that the birds are feeding for themselves. Until now they have been living charmed lives where they have everything delivered. "Once they are free they'll become natural foragers. The real measure of success will be nesting - when they produce their own babies." The cranes have also been given predator avoidance training involving a dog that looks like a fox. Jarret said: "The way we go about training them is to play the alarm calls of adult cranes while walking the dog in the enclosure. "A lot of this is instinct, particularly the calls they make; they just need to know how to respond."

All 21 birds being released have been fitted with small satellite tracking backpacks to monitor where the birds go after their release. A further clutch of eggs will be imported each year, and the process will be repeated until at least 2015 when the WWT hope to have successfully released at least 100 cranes into the English countryside.

# CALL ME DAVE

A team of German and Malay scientists have discovered a new species of chameleon in the rainforests of Madagascar's east coast. The discovery was made near a village that was formerly named Tarzanville, recently renamed Ambodimeloka, so scientists have decided to name the amphibian after the fictional noble savage in order to draw attention to the newly discovered creature and the plight it faces.

Philip-Sebastian Gehring, lead author of the description of the new species, explained: "we

dedicated the new species to the fictional forest man Tarzan in the hope that this famous name will promote awareness and conservation activities for this apparently highly threatened new species and its habitats, Madagascar's mid-altitude rainforest Tarzan stands for a jungle hero and fighting for protecting the forest,"

The new species, *Calumma tarzan,* is 13cm long and has a unique flat snout. The researchers say that the species should be considered critically endangered because of the recent deterioration in the size and quality of habitat.

## SANCTUARY OWNER CAN'T SHELL OUT

Following on from the Three Owls Bird Sanctuary débâcle (see last issue) another animal sanctuary has had to close it's doors to the public. The Tortoise Garden in Sticker, Cornwall, houses more than 400 animals, receives up to 12,000 guests a year and spends £25,000 in caring for their animals.

The sanctuary was forced to close to the public when Cornwall Council felt it had to classify the sanctuary as a zoo because the tortoises were 'wild animals' and not domestic pets. The sanctuary's owner, Joy Bloor was devastated by the council's decision as she would not be able to afford the extra costs a zoo licence would entail. The council's decision seems all the more bizarre when it is considered that all the tortoises in The Tortoise Garden are unwanted or rescued domestic pets.

Bloor hopes that the law can be changed to enable sanctuaries for unwanted pets like hers to still allow members of the public to look around (the sanctuary did not charge entry fees) but has closed to the public for the foreseeable future. Bloor insists none of the tortoises will come to any harm and the sanctuary will continue, just without public access.

## THE NATURE OF BEARDED MONKEY WAS IRREPRESSIBLE

Scientists from the National University of Columbia have discovered a new species of bearded monkey deep in the rainforest. The monkeys were found in the South of Columbia near the border of Peru and have been named *Callicebus caquetensis* and given the local name of Caqueta titi on account of being found in the state of Caqueta.

The bearded monkey is about the size of a domestic cat and has greyish brown hair; the main distinguishing feature from other local monkey species is the lack of a white bar above the eyes and researchers estimate there are less than 250 individuals of the species in the area, which is threatened by logging. The researchers visited the area 30 years after animal behaviour scientist Martin Moynihan reported a sighting of a possible new species of monkey in the area. Researchers were unable to enter the area to confirm the sighting until 2008 because the area was held by rebel groups involved in kidnap and the illegal drugs trade.

The species is more vulnerable than most other monkey species in the area because, according to the researchers, the animals are monogamous and in most cases will produce only one offspring each pair per year, which means that the population will take more time to recover from catastrophic events than similar species that

produce more young, which – in turn - could result in other species encroaching on the monkey's territory. The animals are usually seen moving around the forest in groups of four.

## SEA EAGLE SETBACKS

A Sea Eagle reintroduction programme in Fife, Scotland, has hit a setback after the eagles started to get a taste for chickens. On Thursday the 26th August Treina Samson, daughter of farm owner Alex Samson, came to the chicken coop to find it bereft of poultry: "We shut the chickens in their coop as normal the night before. The next morning, one of the staff went to let them out and saw lots of feathers but no chickens. Then he heard a weird noise, so he shut the coop and looked through the window and saw this huge bird sitting in there. It was quite a shock."

"We are worried that this is stopping our chickens from laying eggs. I don't understand why they decided to release them in Fife. They are amazing birds and it is great to have them in the country - but it is going to have quite an impact on our hens. It's not really the area to be releasing them as it's highly populated and highly agricultural."

Local falconer, Wayne Poole, who released the bird from the coop elaborated: "It had somehow managed to get in through the hatch into the chicken coop. Before they were released into the wild, they fed through a hatch and it's possible they might associate the chicken coop with the hatch and see it as a source of food."

An RSPB spokesperson defended the choice of the area as a site for Sea Eagle reintroduction, saying: "Fife was chosen as a release site because sea eagles are lowland birds. Young birds are inexperienced when it comes to collecting food but they will soon develop the skills to find food in the wild as it is their instinct to do so."

The birds are all fitted with tracking devices and the RSPB will continue to monitor the situation at the farm.

The reintroduction of sea eagles to the east coast of Scotland has been taking place since 2007 with 15 birds being released each year in the Fife area. From next year, 20 chicks will be released on the east coast of Scotland each year. The project follows on from a successful reintroduction programme on the west coast of Scotland and the Western Isles that has been taking place since 1975. The decision was made to start a specific reintroduction programme in the east of Scotland as it would have taken the west coast sea eagles decades to reach the area naturally. The last British native pair bred on the Isle of Skye in 1915, but the species had become locally extinct on the east coast by the mid 19th Century due to persecution.

There is evidence that the sea eagles are spreading farther a-field than the Fife area, with one sea eagle from the programme being spotted in England. The bird took up residence on Staple Island in the Farne Islands to the south-east of Lindisfarne, 5km off the coast of Northumbria on the 27th August. This is only the third time one of the birds has been spotted in England in the last four years and the first time that one appears to have taken up residence this far south. Staple Island is part of the Farne Islands Nature Reserve, which is already a protected site for nesting sea birds.

# THE GIANT RAT OF EAST TIMOR

Australian archaeologists on East Timor have found the remains of a species of giant rat. The rat's bones were found in a remote cave in the small nation. The super rats were three times larger than any rodent currently found on the island. They could have weighed up to six kilograms, and were roughly the same size as a Jack Russell dog.

The rats are thought to have become extinct between 1000-2000 years ago as a result of changes in agricultural practices and habitat destruction. Ken Aplin of the government science agency CSIRO said:

"Large-scale clearing of forest for agriculture probably caused the extinctions, and this may have only been possible following the introduction of metal tools."

"Rodents make up 40 percent of mammalian diversity worldwide and are a key element of ecosystems, important for processes like soil maintenance and seed dispersal."

Aplin believes that there are more undiscovered species of rodent still to be found on the island, including extant species due, in part, to the area's inaccessibility to humans: "Although less than 15 per cent of Timor's original forest cover remains, parts of the island are still heavily forested, so who knows what might be out there?"

# THERE AND GONE

Five new species of frog have been identified among frogs wiped out by the chytridiomycosis fungus epidemic that is currently sweeping through Central America. A team of researchers who noted that 25 of the 63 species of amphibian noted in Panama's Omar Torrijos National Park by surveys up until 2004, had been wiped out by 2008 and decided to undertake DNA bar-coding of the fungi's victims to see if there were any undiscovered species among the dead.

DNA bar-coding is when short genetic sequences that uniquely identify known species are compared to samples from another source to determine if they come from the same species or not. The bar-coding project revealed that there were 11 possible new species of frog among the dead.

"This is the first time that we've used genetic bar-codes - DNA sequences unique to each living organism - to characterise an entire amphibian community," said scientist Eldredge Bermingham, "STRI has also done bar-coding on this scale for tropical trees in our forest dynamics-monitoring plot in Panama. The before-and-after approach we took with the frogs tells us exactly what was lost to this deadly disease -- 33 percent of their evolutionary history."

Karen Lips, associate professor of biology at the University of Maryland, likened the havoc chytridiomycosis was causing to the fire at the library of Alexandria destroying a huge amount of data on how life has coped and evolved with stresses in the past: "When you lose the words, you lose the potential to make new books."

"It's like the extinction of the dinosaurs. The areas where the disease has passed through are like graveyards; there's a void to be filled and we don't know what will happen as a result."

## HORTON, HERE'S A SLENDER LORIS

Scientists from the Zoological Society of London (ZSL) edge of existence programme have successfully filmed the Horton Plains slender loris in Sri-Lanka, thought until recently to be extinct, obtaining the first undeniable evidence of its continued existence since 1937.

The loris has only been seen four times since 1937 and was thought by many to be extinct;

glimpses were often fleeting as on one occasion in 2002 when researchers identified it when their light caught its eyes. The adult male was captured on film after more than 1,000 night time surveys in Sri Lankan forests taken during 200 hours of painstaking work. The researchers were also able to capture 3 live specimens just long enough to measure them and found that they were between 15 and 25cm in length.

ZSL conservation biologist, Dr Craig Turner said: "We are thrilled to have captured the first ever photographs and prove its continued existence - especially after its 65-year disappearing act" Estimates suggest that there are only 100 individual Horton Plains slender loris left in the world but the researchers believe there could be less than 60, making it among the rarest primates in the world.

## FLUTTERING BY

With enthusiastic backing from Sir David Attenborough (president of the charity Butterfly Conservation) and Alan Titchmarsh, MBE (Vice President) a partnership between the charity and Marks and Spencer set up a project whereby members of the public were asked to record how many butterflies they saw in 15 minute time periods between 24th July and 1st August. The sightings were to be submitted online only, and was part of their Plan A commitment to encourage sustainable agriculture and help protect the environment.

From a massive 187,000 individual sightings from all over the country, the top ten commonest species were the small white, large white, gatekeeper, meadow brown, common blue, peacock, green-veined white, red admiral, small tortoiseshell and ringlet. This year's event had more than 10,000 people taking more than 15,000 counts, the project being designed to give an indication to scientists of the state of the nation's butterfly population, particularly in gardens and urban areas.

No day-flying moths made it into the top 10, but

the most common moth reported was the six-spot burnet (which came in at 13th place overall with just under 4,000 individual moths counted). This was followed by the silver Y and then, some way behind, the hummingbird hawkmoth.

However, there were interesting results for the butterflies, which included an impressive number of sightings of the small tortoiseshell, which in recent years has seen a decline of 82% in numbers in southeast England. This may be due to the parasitic fly *Sturmia bella,* but the recent count results indicate strong signs of recovery. This butterfly was the ninth commonest seen across the UK, doing even better in garden habitats. Conservationists were very pleased with the gatekeeper's results as this orange butterfly has endured a run of very bad years. During *this* count, however, it reached number three in the top ten and was seen in greatest numbers in fields and other rural habitats. It was also found to be thriving in major cities such as London and Manchester.

However, scientists at the charity have warned that, despite the pleasing results for some species, most butterflies continue to face a serious long-term decline. Butterflies and moths act as important indicators that alert us to problems with the environment, and if *their* numbers are falling, it is a strong indication that other wildlife is also in decline.

Many butterfly species will be affected by loss of habitats such as flower-rich grassland, and their future will be marred by the intensification of farming methods. There are also problems in woodland habitats caused by lack of management and Butterfly Conservation is working with landowners and other conservation organisations to help reverse such declines as a matter of urgency.

Butterfly Conservation's Surveys Manager, Richard Fox, said: "A big thank you to all who have made the big butterfly count the biggest ever weekly count of butterflies anywhere in the world! We were impressed by the Gatekeeper and delighted to see the Small Tortoise-

shell in the top ten as it had become a scarce sight, particularly in the south. It's been a fantastic start and the big butterfly count will continue in 2011. With the public's help, we'll be able to compare how butterflies and moths have fared. We hope people from all over the UK will help us take the pulse of nature in years to come."

The full results of the count can be viewed online at www.bigbutterflycount.org and to find out more about the decline of butterflies and how you can help, visit www.butterfly-conservation.org. In 2003 Butterfly Conversation partnered with the Forestry Commision and Forest Research in Scotland, and the Allt Mhuic nature reserve was set up on the north shore of Loch Arkaig. This aimed to support several species of butterfly, in particular the nationally important chequered skipper, which only occurs in the UK within a 25 miles radius of Fort William. In August this year, the butterfly reserve reported a significant increase in butterfly numbers and the intervention of Highland cattle could be the reason. As part of the project the Commission utilised 15 young Highland cattle in a large scale habitat and species management programme, but in 2009 it was apparent that this conservation grazing programme was not working as was hoped and the butterfly numbers were still in decline.

Kenneth Knott, for the Commission, said: "The decision was taken to change the grazing regime on the lower slopes from summer to winter grazing and early indications suggest that the change is benefiting butterfly numbers. As the season has progressed at least five species on the site have shown improvements in numbers, which is excellent news and we are looking forward to getting the end of season figures to compare to previous years."

There has been a 160% increase (up from 3 to 8 in the monitored transect) of the chequered skipper butterfly and the first ever male territories have been mapped. Indications are that numbers of the dark green fritillary have shown a 250% (up from 8 to 28) increase while the small pearl bordered fritillary and the speckled wood butterflies have doubled in numbers.

The local recorder on the site Tony Millard said: "The whole site looks really good and the results this year are fantastic. The introduction of the cattle has made a real difference to the quality of the site, making the conditions and the habitat much more amenable to several butterfly species. An early success like this is really encouraging and we hope we are on course to see further improvement next year. It's a great example of how more natural land management techniques can bring additional benefits."

## BATFISH INSANE

Two new species of pancake batfish have been identified in the Gulf of Mexico. The batfish were all previously thought to be one species, *Halieutichthys aculeatus*, but researchers, including John Sparks from the American Museum of Natural History, New York, determined that there were two additional species that had been incorrectly identified as *H. aculeatus* from the same area. The new species have been named *H. intermedius* and *H. bispinosus*.

All three fish are particularly imperilled by the recent oil spill in the Gulf as they live in waters directly affected by the slick, with one of the newly discovered species and *H. aculeaus* only documented in affected waters. "If we are still finding new species of fishes in the Gulf, imagine how much diversity, especially microdiversity, is out there that we do not know about." Sparks said. "These discoveries underscore the potential loss of undocumented biodiversity that a disaster of this scale may portend."

Pancake batfish have round, flat bodies with large heads and mouths that they thrust forward to take prey and a dorsal fin that exudes fluid that acts like a bait to prey. They move along the sea floor using arm-like fins in a manner evocative of a bat walking.

# BRITISH LIVEBEARER ASSOCIATION

Anyone interested in keeping and maintaining livebearing fish, whether wildform or cultivated, would do well to check out the British Livebearer Association. They were set up in order to keep members informed of the availability and identification of livebearers, along with husbandry information. They also give tips on general fishkeeping and report on any new developments in conservation as well as give descriptions of new species and any name changes that may crop up – mainly via a quarterly magazine, which members receive around March, June, September and December. Members can opt for a paper or an e-form magazine.

The Association also hold a minimum of two events a year – a spring auction and mini-convention in March and an auction, fish show and convention around October. Sometimes other events are held during the year and details are published in the magazine.

If you are interested in guppies, then there is also a sub-group that holds other events during the year.

The British Livebearer Association also runs a species control and maintenance programme. This is to facilitate the availability of new species throughout the membership. There is also a breeder's award programme that rewards commitment and success in breeding wild species of livebearer fish and members are encouraged to take part in these programmes.

To find out more about the Association, and to become a member, check out their website at: http://www.britishlivebearerassociation.co.uk/

# TWINNING DAY 2010

On Sunday 15th August the majestic Whistler Room at Pickering Memorial Hall played host to a special Twinning Day involving the members and friends of the Ryedale Aquarist Society and STAMPS.

The day began with a special presentation on the subject of Pencilfish that was given by Miss Amy and Mrs. Wendy Charters.

Special guest for the event was Mr. Steve Dent of the Yorkshire Cichlid Group who presented two talks based around his favourite cichlids of South and Central America. The people present enjoyed taking part in a food and drink quiz that was devised by Miss Sue Marshall.

Y.A.A.S. 'A' Class Judge Mr. Trevor Douglas judged the 9-class mini-show in which a total of 44 fish were entered. The quality of fish was excellent and included *Corydoras pantanalensis*, Chinese Roundtail Guppy, Schubert's Golden Barb and C6.

A sales table was also held with various cichlids, livebearers and fish foods causing much interest. The Aquarium Gazette CD magazine was also present with an information stand.

Finally, Twinning Day would not be Twinning Day without our traditional 'pot luck lunch'. A big thanks to all who brought along such a wonderful selection of sweet and savoury

# CLUB NEWS

items!

A big thank you to all the people present for all the work that was done on this very special day and for friendship.

This very special photograph, featuring Bede Kerrigan and Frank Tolomeo, sums up what Twinning Day is all about.

Looking forward, the Ryedale A.S. Open Show 2011 will be held on Sunday 24th April at Old Malton Memorial Hall, North Yorkshire.

# THE BRITISH HERPETOLOGICAL SOCIETY

The British Herpetological Society [BHS] is a registered charity run by voluntary workers and was established in 1947. It has grown into one of the most prestigious of its kind in the world with over 600 members.

The society produces three regular publications:

- A monthly newsletter called The Natter Jack which contains news items, useful information about meetings and events, plus a small classified section.

- The quarterly Herpetological Bulletin with a wide range of articles related to the subject, which include full-length papers, book reviews, letters, society news etc. The emphasis of this bulletin is on natural history, captive breeding, care, plus veterinary and behavioural matters.

- The quarterly scientific Herpetological Journal which is ranked as one of the leading scientific journals dedicated to the subject.

All the above are available to members, and if you are interested in becoming a member, you should check out their website at: http://www.thebhs.org or write to them at: The British Herpetological Society, 11 Strathmore Place, Montrose, Angus, DD10 8LQ.

The Society also runs a Young Herpetologists Club for those aged 5 – 17 years of age, which was established in 1980. This encourages those youngsters who are interested in the subject to learn more about amphibians and reptiles.

The BHS is also one of the few such societies to possess a library, and it has many books, and reprints, some of which can be borrowed by members. The library holds the Corkhill bequest, which contains many rare books on venomous snakes and religion and snakes.

# KEEPING LOCUSTS AS PETS

## CORINNA DOWNES

If you think of locusts you most likely bring to mind two things: swarms that devastate crops and if you are a herp keeper, feeder food – they are more nutritious than crickets or mealworms, and their large size and attractive yellow abdomen make them an appealing treat to reptiles and larger invertebrates.

The locust is actually the swarming phase of the short-horned grasshoppers of the family Acrididae. Derived from the Vulgar Latin *locusta*, the word 'locust' was originally used to refer to various types of crustaceans and insects.

Two of the most commonly available are the desert locust and the migratory locust.

The desert locust (*Schistocerca gregaria*) has been a centuries-old threat to agriculture in Africa, the Middle East and Asia, and these insects can affect the livelihood of at least one tenth of the world's human population in their voraciousness, being probably the most dangerous of locusts as it has the habit of swarming and flying rapidly across great distances. They can live between three and six months and lead a solitary life until it rains, when the female lays her eggs in sandy soil and helped with the vegetation that grows as a result of the rain, the young are provided with shelter and food as they develop into winged adults. The

desert locust has two to five generations per year; thus great numbers are produced – there is a ten- to sixteen-fold increase in numbers from one generation to the next.

The migratory locust (*Locusta migratoria*) is the most widespread locust species and is found throughout Africa, Asia, Australia and New Zealand. It was once common in Europe but has now become rare there.

Locust young are called hoppers or nymphs and when the vegetation grows in such a way to cause them to congregate to feed (if there has been enough rain for most to hatch) the close contact causes their hind legs to bump against each other. This causes metabolic and behavioural changes and they transform from solitary to gregarious. When this stage is reached they then change from green to yellow and black, and the adults change from brown to red for immature and yellow for mature. Their bodies become shorter and they become attracted to each other because of the pheromones emitted, which in turn encourages hopper bands and subsequently swarm formations.

When swarming, these locusts fly with and at roughly the same speed as the wind and can cover 100 to 200 km a day. They can fly up to around 2,000 metres above sea level before it becomes too cold; thus they cannot cross tall mountain ranges. They can eat the rough equivalent of their body mass each day, which works out at around two grams of green vegetation. For every million locusts, one tonne of food is eaten.

However, locusts can be kept as pets. Some may be concerned about letting loose a swarm of eating-machines should some escape, but rest assured that they would not behave as their wild cousins would.

## HABITAT

The cage itself need not be too big – it should allow the locusts plenty of climbing space and allow them to warm themselves by the lamp and to move to a cooler place to look for food. A couple of pairs of adult locusts will be happy in a cage of 28 x 16 x 10cm. If a wooden cage is used, the heat mat should be used to cover most of the rear wall (taped on the inside) or one on either side if the cage is glass or plastic. When taping, ensure that the edges are taped firmly down as the insects may well nibble at a free edge and any exposed adhesive will act like flypaper, which, of course, could prove lethal.

Add plenty of twigs for them to climb on, and as they may also wish to hide themselves and hang protected when shedding, some care should be taken to ensure a suitably secluded spot is provided.

Locusts like a very dry atmosphere with a good deal of ventilation. Fresh air circulating in the cage is essential, so use the upward draughts caused by heated air rising to flush out the cage and drag fresh air in.

Locusts thrive under bright light so artificial light is a primary addition. They also need radiant heat so both can be supplied by a basking spot-lamp. The temperature at the hottest part of the cage should be between 34 and 36°C and should be maintained for about ten hours each day.

The heat mat is used to provide background heat and this can be left on all the time. The temperature should not fall below 20°C. If the cage becomes too hot put it in a cooler place, but if it is not high enough, a higher wattage spot-lamp will have to be utilised.

The cage will need cleaning about once a week as faeces and debris will build up – a good all-round cleaner would be Vetaclene – followed by a thorough rinse and dry. Other materials should be replaced when soiled.

## FEEDING

They will eat almost anything in the plant line – they are not fussy. They like fresh leafy vegetables – cabbage and kale are the best of those that can be bought from your local fruit and veg store. However, lettuces are not great as they are too watery. When selecting your veg, go for the loose leaf varieties rather than the tightly packed kind. They also like cereals and grasses. The best thing to do is experiment with anything edible to see if their appetites are aroused! However, caution is obviously called for if feeding from the garden and places where insecticides may have been used. Feed them fresh food every day. You can also purchase pre-made formula, which contains various mixed up foods, ground into a powder.

Some dry bran and hay are excellent too as they will not deteriorate in the dry heat of the cage and can be left in the cage indefinitely. Locusts will also eat cornflakes, oats and biscuits. No provision of water is necessary as they get all they need from the fresh vegetation. Water may in fact cause problems – dampness or even slightly raised humidity within the cage must be avoided as damp conditions can cause the locusts to die of mould or fungal infection.

Locusts are gregarious and look best in groups. If they are kept on their own – especially during the early stages – they develop into what looks like an entirely different animal. This is known as the solitary phase as noted above, and more often than not the creature takes the form of any other large green grasshopper.

Locusts do not like being handled so if it is necessary to do so, the best way is to pick them up in an enclosed hand. Apart from fleeing, they have two defences: one is to vomit a fluid that will stain your hand, and the other is to kick out with their spiky legs and feet. This will feel like tiny pin pricks, but neither defence is dangerous.

## BREEDING

Sexing locusts is fairly easy. Upon inspection of the tail-end of the abdomen males will have slightly larger cerci or small feelers. Sexes are also easy distinguishable in adults as the males are yellow and the females are a buff/beige colour. Breeding is not difficult – mature males and females will mate easily if in the right environment. The female will lay an egg pod (ootheca) containing several dozen eggs, which is buried in a damp medium – a dish or bowl of moist sand or peat should be provided and be about four to five inches deep for her to lay eggs in. The medium must be kept moist so that when she lays her eggs, the tunnel will not collapse. When ready, she will insert her abdomen into the sand, and lay her eggs either deep under or just below the surface. Baby locusts will emerge after around ten to fifteen days, and can be fed on the same foods as the adults were previously feeding on.

Several weeks may lapse before locusts reach adulthood, and this will largely depend upon the temperature, food availability and stocking density. Once they do reach adulthood they only live for a few more weeks. They do not suffer from many diseases and special attention is rarely needed. Vetadine can be used as first aid treatment for any wounded locust, by dabbing the wound with it.

Pet locusts do not pose a real threat to human health although an allergic reaction may occur with sensitive individuals, which will cease once contact is broken off. As with any other animal, normal hygiene when handling and caring for them should be observed.

## SUMMARY

If you are attracted to these infamous insects rather than in just using them as feeder food, then keeping them as pets may prove an interesting and somewhat different hobby.

## RICHARD MUIRHEAD

Locusts have turned up in Britain from time to time, brought in by winds from the Black Sea area or North Africa, as testified by a number of naturalists and Charles Fort in *Lo!*. According to Williams: 'The earliest reference to an occurrence in this country [Britain] was in the fourteenth century. In October 1693 two "vast swarms" were seen in North Wales, as a number of scattered individuals in South Wales'.

Between the period of 1900 and 1940 migratory locusts were recorded in ten of the years, some as far north as the Orkney and Shetland Islands. Usually only one or two individuals were noted in a year, but in 1931 numerous records were scattered over eastern England as far north as Yorkshire. From a careful study of the measurements of specimens caught here, it appears certain that nearly all of them came from the Black Sea area, after a cross-country flight of about 1,600 miles in a straight line. Only adult locusts have been found in Britain, but Miss Waloff - who has summarised the information - says 'that they may arrive sexually mature, and

it is possible that during a hot summer they might lay eggs in south-eastern England. But it is very unlikely that any would survive our winter.' [1] Of course with global warming this last point may change as we are seeing many exotic species turning up in the UK every year now.

'In October 1869, a number [of desert locusts, *S.gregaria*] were found 'scattered over southwest England from Plymouth in the south to Derby and Nottingham in the north.....This was the first record of the desert locust anywhere in Europe.'[2].

In November of 1869 there was a report of a locust from Truro, Cornwall - fortunately with a good description of its appearance: 'Locust - On the 11th November, in a bright gleam of sunshine, we caught a locust on the jessamine growing over the front of our house.........I presume it is a migratory locust; but how is it that there is no green colour anywhere? There is a little bit of red on the body; but generally it is of a flat, grey hue. It is about 2 ½ inches in length; that is, the body without the legs; and the spread of the wings from tip to tip, nearly 5 inches. I hear that several have been caught in this neighbourhood during the last few weeks…' [3]

Fort reported: 'In the summer of 1921, England was bereft of insects. The destruction of insects, in England, by the drought of 1921, was, very likely, unequalled at any other time, anyway for a century or more…..Then came clouds of insects and plagues of insects……..Locusts appeared (London *Weekly Dispatch,* July 6) [4]

But returning to 1869: 'Upon the 8th and 9th of October, locusts appeared in large numbers, in some places, in Pembrokeshire, Derbyshire, Gloucestershire, and Cornwall' [5] says Fort, also noting an invasion of ladybirds.

The desert locust did not turn up again until October 1954 'when on the afternoon of 17th October, two were captured in the Scilly Islands, and one just off the shore at Tranmore in Co.Waterford, Eire.' [6] In October and November 1988 more than 70 desert locusts turned up in southern England.[7] Jumping ahead to April 1994, J.Aynsley saw a desert locust on April 11th 1994 at Common Moor Post Office garden near Liskeard, Cornwall. [8].

No examples of British locusts have been found after 1994 and editors of *Fortean Times* - Bob Rickard and Paul Sieveking - could not enlighten me further as to whether or not there were any. However, for those of you who want to research further the 'General Index' to *Fortean Times* issues 1-66 November 1973 - December 1992 mentions locusts in Surrey no. 40 p.6 and UK generally in 56 p.16. The comprehensive indices in *Fortean Studies* volumes 1-7 contained nothing of relevance.

1. C.B. Williams  Insect Migration. (London: Collins,1971) p.85
2. C.B.Williams Ibid p.87
3. H.Budge `Locust` in The Naturalist`s Note-Book (1870) p.44
4. C.Fort The Complete Books of Charles Fort - Lo! (New York: Dover Publications, Inc, 1974) p.740
5. C.Fort Ibid p.747
6. C.B Williams op.cit p.87
7. J.A. Marshall and E.C.M.Haes *Grasshoppers and allied insects of Great Britain and Ireland* (Colchester: Harley,1988) pp 145-147
8. National Biodiversity Network  http://data.nbn.org.uk/ [accessed 09/09/2010]

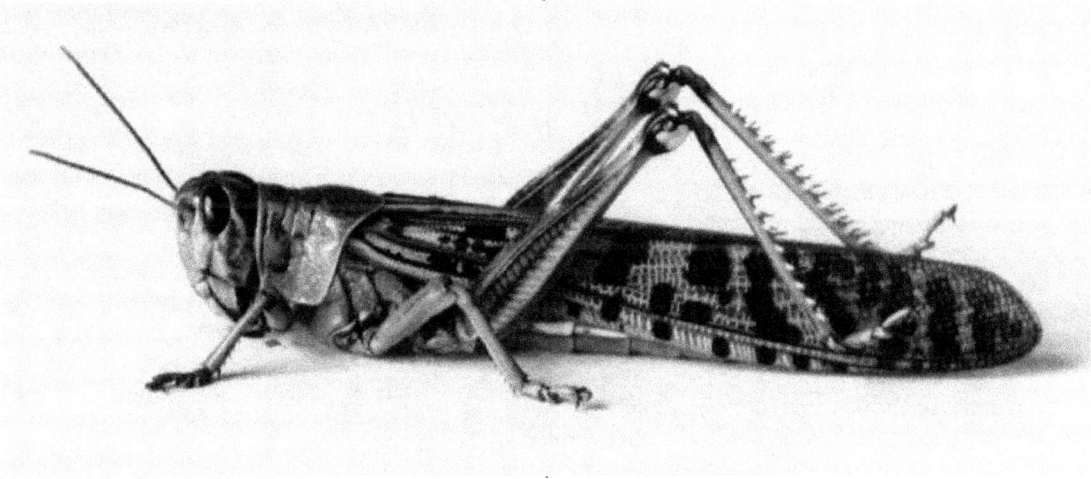

# A JEKYLL AND HYDE OF CHARACINS

# By David Marshall

## BACKGROUND

Many South American characins live 'Jekyll and Hyde' lives. From an aquarist's point of view this is mostly seen through the medium-sized group of characins known commonly as Headstanders. The wonderful looking Abramites Headstander (*Abramites hypselonotus*), so resplendent in its marbled body pattern, has the devious mind, and capability, for not only stripping the fins from similar-sized aquarium fish but also for delivering deadly nips to the bodies of those companions smaller than itself. However, for the purpose of our article, we will concentrate on a much smaller characin that you may well be familiar with. Under the common name of Buenos Aires tetra, this character fits this 'double life' profile to a tee.

Although the scientific name of *Hemigrammus caudovittatus* continues to be used in connec-

tion with this species it has now been transferred, or returned as the case may be, to the well-known characin genus *Hyphessobrycon,* where it can be found as *Hyphessobrycon anisitsi*. Although Buenos Aires tetra is the universally accepted common name you may have also see this fish offered for sale as the diamond spot characin, the red cross fish or simply flag tetra (which can cause confusion with its cousins *Hyphessobrycon heterorhabidus* and *Semaprochilodus theraponura* that are also known under this common name).

I began writing this article believing that the Buenos Aires reference pointed towards Argentina as the endemic homeland of our subject specie and this belief was bolstered by the fact that the first aquarium specimens were reported as exports from the river Plate basin. However, my research into *Hyphessobrycon anisitsi* also uncovered populations that occur in areas of Brazil, Uruguay, Guyana and Paraguay. I was surprised to learn that, in the early 1970s, a feral population was reported as thriving on the

Philippines, but the current state of this population is unclear.

Although found in a variety of natural habitats, these shoaling fish are most at home in slow moving ponds containing copious amounts of natural vegetation. Not only does this vegetation provide natural cover, but it is also forms an important part of their diet. Exactly what nutritional benefits the Buenos Aires tetras derive from this particular food source has yet to be scientifically studied. Aquatic worms, insects and crustaceans account for their protein needs.

## THE JEKYLL

The Buenos Aires tetra, which has large eyes and the familiar bullet shaped body of many of the smaller tetra species, is available in both natural and albino colour forms. When kept in the right conditions, with a little natural light entering the aquarium, the colours of both varieties are a true joy to behold and, thus, bring hours of pleasure to the human eye.

The background body colouration of the natural form is brownish-yellow. The back is pale green or brown whilst the flanks boast a metallic sheen that compliments the blue-green stomach area. A black lateral line runs across the top of the stomach and into the caudal peduncle. Above the lateral line runs a second line of neon green. At the caudal peduncle (where the body ends and the tail begins) you will find a diamond-shaped (hence the alternate common name) black spot that has small, and often faint, yellow or red pyramid markings at its rear end. In fish which are in good condition a brass coloured line, reminiscent of a breaking wave and varying in size, can be seen on the top of the body and runs from the gill covers to a point just short of the caudal peduncle. The dorsal and caudal fins are yellow with red overtones at their base, whilst the anal fin is a bright mix of brown and red.

Although it has the characteristic pink eyes the

albino variety, which was developed by Far Eastern commercial aquarium fish breeders, it always appears to be more of a xanthic nature rather than a true albino. Here the black lateral line is swapped for one of deep red. Amazingly, the green neon line (something you often see in albino ruby sharks) is retained. There are changes also within the fins, as these lose their yellow to become deep red.

With both forms the maximum body size attained is 4". On the whole males tend to be slimmer and have much brighter body colours than their female counterparts.

## THE HYDE

The first time I came across the Buenos Aires tetra my 'old friend' Mr. Norman Bird, who owned a North Yorkshire aquatic retail outlet for many years, described them to me as 'mini-beasts'. Thirty years of experience has not caused me to alter his description in any way!

The Buenos Aires tetra is a naturally shoaling fish so for an aquarium population we look towards the purchase of six or more individuals. You can mix the two colour forms but I have witnessed instances in which the natural form will see the albino as a totally different species and thus will try and form a position of dominance over them. Within a group of either colour form, a distinct feeding time 'pecking order' will always emerge.

As we have already mentioned South American headstanders are well-known in the aquarium hobby for the sheer joy they show in nipping at the fins of other fish species. Well, sadly, this is also the case with the Buenos Aires tetra, which can leave its poor victims in an even worse state. This means that we should never house them alongside any fish that have veiled fins or are of a nervous nature. If they feel they can dominate a situation they will!

Unfortunately, it can get worse! As we mentioned in our Issue 8 article the African short-

nosed clown tetra (*Distichodus sexfasciatus*) has a reputation for seeking out a particular fish that it forms a dislike to and then beginning a campaign of hounding and biting against its unfortunate victim. It is not unknown for an individual Buenos Aires tetra to do the same, and once the other shoal members sense the 'buzz,' the attentions of the group can turn towards the thought of eliminating such a victim from their aquarium.

Are these mini-beasts best kept alone? Although I have succesfully kept *Synodontis* and various driftwood catfish as companions I recommend that Buenos Aires tetras be kept in a single species aquarium. However, even then, there can be unexpected problems. For quite some time a large shoal of mature Buenos Aires tetra had lived happily together, in my care, without any trouble whatsoever. When a fellow 'fish fanatic' visited we decided to remove three of the shoal, in order to choose one for the showing bench. With our job done all three, roughly thirty minutes after their initial

removal, were returned. To our horror, the other group members treated them as strangers and attacked in a way their *Abramites* cousin would have been proud of. The three fish were quickly removed to new quarters and never returned to their old home.

If you have not kept the Buenos Aires tetra before this section was not written to put you off doing so, but just to warn you about aspects of their natural behaviour. If your shoal settles down with each other they will bring much pleasure with their beauty and active attitude towards life.

## AQUARIUM CARE

We begin this section with a warning. When we bring our Buenos Aires tetras home from the retail outlet great care has to be taken when releasing them into their new aquarium because, like all characins, they are very susceptible to going into shock at this time. To try and avoid such shock we float their transportation bag for at least ten minutes and then mix water from the bag with that of the aquarium for several further minutes before releasing the fish into their new quarters.

These fish require plenty of space in which to dart around so the minimum sized aquarium in which to house a small shoal is one of 36x12x12". To highlight the body colouration of our charges we employ darkly coloured fine-grained gravel as a substratum. This is enhanced by placing pieces of coal, of various sizes, around the aquarium base. In the middle of our aquarium we place a large piece of matured bogwood with smaller pieces to each side. If we are to keep any of the bottom dwelling fish mentioned above in our set-up, we add halved ceramic plant pots and broken coconut shells etc. at this stage.

Very few real plants will survive the natural attention of a Buenos Aires tetra. However, the sturdiness of Amazon swords (*Echinodorus* species) means that these can be used for background planting, whereas the toughness of the fronds of Java moss (*Vesicularia dubyana*) make it too hard going for prolonged attention so spherical bunches of this plant are fastened to the wood as a foreground feature. Although the purists among our readers will cry 'shame on you', a variety of plastic plants fill the mid-section. What we must not do is fully plant the aquarium, as our fish require some open swimming space. It is not uncommon for Buenos Aires tetras to actually rest among the leaves of aquatic plants so if you find one in such a position do not panic that it is not well etc.

As far as the pH of their water is concerned the Buenos Aires tetra will be happy at any point in the range of 5.8 to 8.5. If our aim is to eventually breed our fish, we need to maintain them, on their own, at a pH no greater than 6.5. Although for breeding purposes, we can slowly condition Buenos Aires tetras to move down from a higher pH to one between 5.8 and 6.5 through using peat or a commercial water softener, this is extremely risky.

On several occasions I have seen Buenos Aires tetras for sale in the coldwater section of aquatic retail outlets. Although they can be kept at room temperature, due to the sub-tropical climate in which some populations occur, I believe this to be too cold for their welfare so I advise that these fish be kept at a temperature of around 74°F. Provided that you make plenty of small regular water changes, which they love, filtration can be kept to a minimum.

An important point often overlooked with regard to our subject species is that these fish do have the ability and agility to leap clear of their aquarium so we must always employ a tightly fitting glass cover or condensation shield. If you have never kept these fish before you will be surprised by their turn of speed and activity!

Buenos Aires tetras are 'Koi of the characin world' as they literally beg to be fed at every opportunity. This begging would appear to be based on metabolic needs, as their hectic activities may burn up energy reserves pretty quickly, so to cater for their hunger they are best given two or three small feedings a day. They love granulated foods, frozen bloodworms, vegetable-based flakes, peas, shredded sprout and lettuce.

Although very hardy looking, these fish are susceptible to a number of the more common fish illnesses seen within aquarium circles thus, to try and avoid such problems, always maintain a good schedule of aquarium maintenance.

## BREEDING

We need two more aquariums if we are to make a spawning attempt. The first of these needs to be 36x12x12" and must be placed in a position to receive plenty of morning sunshine. This aquarium must be sterilised with a strong salt solution. Before any fish are added the solution has to be thoroughly washed away. The second aquarium can be 24x12x12" and is prepared as per our community set-up. A banana worm culture as well as brine shrimp eggs and hatching equipment are a must.

The 36x12x12" now occupies our time. As spawning aids, floating plants, bunches of bound *Elodea* and well-washed woollen spawning maps are added. Around the bases of the bound plants we make a loose coal or slate square. Between the square and the plant bases we add as many marbles as we can. We add a mixture of rain and tap water.

As we have already mentioned any fish we hope to use for a spawning attempt should, ideally, not have been kept at a pH greater than 6.5. Several days before the breeding aquarium is set-up we condition the tetras in our community. The key element is to change from frozen to living bloodworms and to feed this in large amounts. From watching the group activity in our community we should have a good idea as to which two fish have formed a compatible pair. The male will be more colourful and have a slimmer body profile than the female, who should be filled out with roe.

With the breeding aquarium looking established, and of a matching pH to our community aquarium, the female is caught and moved over to the spawning aquarium where she is left on her own and well fed for a couple of days, while we slightly increase the temperature of the water around her. The male, who is floated in a jam jar, is then added during afternoon hours. There is much written in aquatic literature, about how, in the evening, we change around a quarter of the aquarium water but this is too risky and can greatly alter the pH, no matter how much care we take.

There is no guarantee that a successful spawning will immediately follow and we may have to be patient for a few days, make more water changes or even swap the partners around. Hopefully, though, this will not be the case and the morning sun will trigger off a spawning attempt. When this happens the male vigorously drives the female first towards the plants and then towards the substrate. As the female

sheds her eggs the male quickly fertilises them. No parental care is given and one or both of the pair may break off from spawning to snatch small numbers of their own caviar.

There is no cast iron guarantee that the pair, in particular the female, will emerge from such a spawning unscathed. Once we realise spawning is over the pair are removed to the 24x12x12" aquarium. Hopefully, after a rest, they can return to the community but, as we have seen, this is not always possible.

Sadly ignored these days is the '70's golden rule' for successfully hatching characin eggs. This involves shading the eggs from light by covering the aquarium with newspaper or a blanket and I recommend that you follow this step. Hopefully, we will have close to 200 eggs (150 is the average number for small South American characins), which are semi-adhesive, collected among the plants, in the mops, between the marbles and upon the substrate. The eggs are quick to hatch, usually between 20 and 24 hours after fertilisation, and the fry, which resemble tiny glass splinters, cling to the plants etc.

Using a magnifying glass we watch for signs that the fry have become free swimming, by which time we have removed the newspaper or blanket cover, and, with their yolk sacs used up, we begin feeding them by using green water from our established community aquarium. This provides the fry with enough nutrients for a couple of days before we move on to using a commercial fry food. Small water changes are made and great care has to be taken, as we do not wish to either over feed the fry or pollute their aquarium.

A few days down the line we add live brine shrimps for the first time. If the shrimps do not seem to be taken we revert to feeding banana-worm. Up until this point any fry growth would have been tiny, and even hard to detect, but the shrimps in particular should help speed this process along. Our aim is to raise a small number of strong fish. If all goes according to plan we will have seen the first colours upon the bodies of our young fish (this often begins with colour in the dorsal fin) within twelve weeks of their initial hatching and at this time they should be just under a body size of 1". Perhaps the only drawback is that their adult traits can start to emerge at this time.

## Summary

*Hyphessobrycon anisitsi* is not a characin for the faint hearted. Within its nature are many good and bad traits. However, given the correct care, these medium-sized tetras will thrive in captivity.

THE CENTRE FOR FORTEAN ZOOLOGY
www.cfz.org.uk

WHICH SIDE ARE YOU ON?

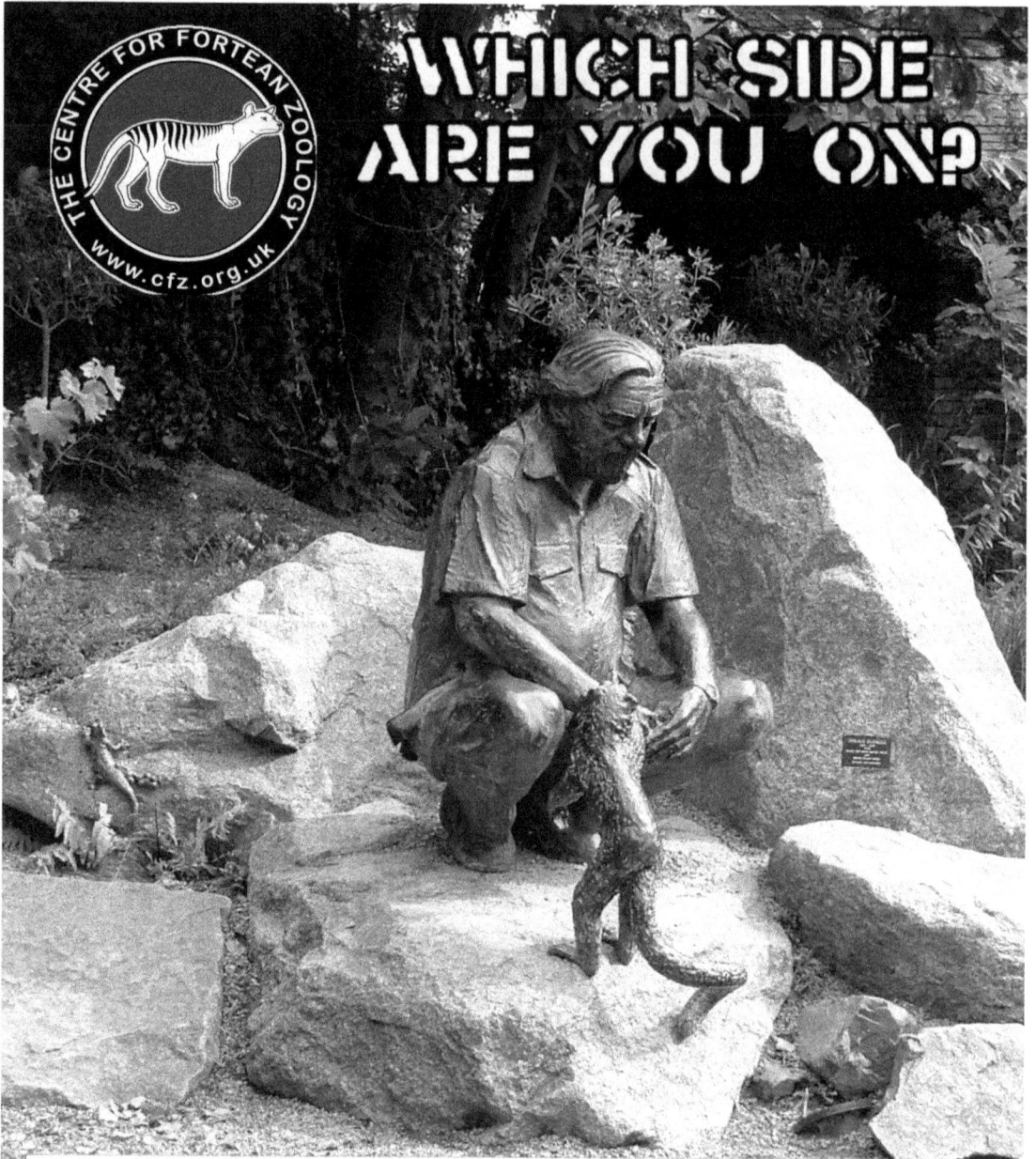

OUR HERO GERALD DURRELL (1925-1995) WAS THE FATHER OF MODERN CONSERVATION. HE BELIEVED THAT CHILDREN SHOULD GROW UP SURROUNDED BY ANIMALS AND BOOKS.

WE AGREE WITH HIM.

BUT IT SEEMS THAT OTHERS DON'T!

CFZ OUTREACH: Educators with Attitude

# HAWKMOTHS AND TIGERS AND THE BUTTERFLY EFFECT

**Jonathan Downes**

Although I have been notionally a cryptozoologist for many years, one of the things that has intrigued me most has always been what could be described as microcryptozoology: that is cryptozoology on a very small scale dealing with the mystery fauna of relatively small areas of the earth. A primary example of this is the study which has engaged Richard Muirhead and myself on and off since we were children; the study of the mystery animals of Hong Kong. Hong Kong is a relatively small slice of China consisting of 426 square miles, carried over a peninsula and a series of several hundred islands to the east of the Pearl River delta on the coast of the South China Sea. Between 1851 and 1997 it was under British rule and, as natural history was the favoured pursuit of the British middle and upper classes for much of this time, it is not surprising that the fauna and flora of the territory were chronicled meticulously.

Both Richard and I grew up in Hong Kong under British rule and our single 'sacred texts' included the *Hong Kong Countryside* (1951) by G.A.C. Herklots, *Hong Kong Birds* (1953) by the same author, *Hong Kong Butterflies* (1960) by Major J.C.S. Marsh, and *Hong Kong Mammals* (1967) by Patricia Marshall. Since then there have been successive waves of new books on the subjects with each successive decade.

And the strangest thing about this is that each of these successive waves of reference literature gives a significantly different list of animal residents in the former British colony.

This isn't – as you might be forgiven for thinking – because certain species have become extinct. Not always at any rate. OK, there hasn't been a confirmed tiger sighting since 1947, a leopard sighting since 1957 or a Chinese red fox that was not introduced by man since the mid-1950s. The large Indian civet (*Viverra zibetha*) became listed as extinct some time in the 1980s. But despite these sad losses, the aetiology of the changing fauna of Hong Kong is far more complicated than merely the slow, inextricable decline of most species that are unable to compete with increasing pollution and urbanisation.

Writing in the '60s for example, Patricia Marshall noted that there had been no crab-eating mongooses seen since the early 1950s, and for the next twenty years they were considered to be extinct in the area. Then in the mid- to late-1980s, not only was the crab-eating mongoose (*Herpestes urva*) rediscovered, and furthermore rediscovered in quite considerable numbers, but a second species of mongoose - the small Asian mongoose (*H. javanicus*) - was discovered living alongside it. Within a few years, two more species of small carnivore - the yellow-throated marten (*Martes flavigula)* and the yellow bellied weasel (*Mustela kathiah)* - had been recorded for the very first time, and the Chinese otter (*Lutra lutra chinensis)* which had been considered to be extinct in the territory for decades was making a remarkable comeback. There were even sightings suspected to be of the local sub-species of red fox (*Vulpes vulpes hoole)* on Lantau Island, where there hadn't been reports in living memory.

I find these fluctuations in fauna absolutely fascinating, and the study of such things has taken up much of my time in recent years. I find an irresistible parallel between them and the conceptual science of psychohistory as mooted by science fiction author Isaac Asimov, and Hong Kong, which has undergone so many socio-political changes in the last century, having been ruled by imperial Britain, post-imperial Britain, imperial Japan, a modicum of home rule, and now communist China under the "one country two systems" mandate agreed in the early 1980s between Margaret Thatcher and Deng Xioping, is a perfect living laboratory for the study of such things.

But these changes (apparent or otherwise) in the biodiversity of a region can be seen everywhere you go. Recently, news releases from English Nature have confirmed that pine martens are not, and never have been, extinct in England – a supposition which I first published in 1994 in the first issue of *Animals & Men.* If *M. martes* – a lively carnivore the size of a cat – can (like Wordsworth's *Lucy*) still dwell amongst the untrodden ways, then when one is beginning to look at the spread of invertebrate species, especially nocturnal invertebrate species, one can see that all sorts of possibilities can and do exist.

For example, one of the most sought after day flying moths in the UK is *Euplagia quadpunctaria,* the Jersey tiger moth, which historically has only ever existed in a number of distinct and isolated areas of the south coast. I have been lucky enough to live in three of these; the nurses' home at Langdon Hospital just outside

Dawlish, the small village of Star Cross on the A379 between Torquay and Exeter, and Exwick – an unprepossessing suburb of Exeter where the CFZ and I lived for 20 years. I can still remember the amazing adrenalin rush that I experienced when I first saw one of these beautiful day flying moths. It was so rare, so fantastic a creature that my heart nearly stopped in delight.

For well over a hundred years it has been suspected that the reason that this species first appeared in any numbers in Britain, first at Star Cross and then in the suburbs of Exeter, was because of skullduggery from an otherwise blameless Church of England vicar – the Reverend Francis Orpen Morris – who it was suspected by many entomological historians, including the irrepressible P.B.M. Allan writing in 1948 in his book *Moths and Memories* had introduced this delightful little insect to the region for reasons of his own, probably involving the chance of selling some high-priced specimens on the increasingly lucrative collectors' market. Accusations and counter-accusations flew across the pages of the technical literature and even such luminaries as Edward Newman, the doyen of British Lepidoptera at the time, were accused of shady dealings.

Today, the 3rd September 2010, the 71st anniversary of Britain entering World War II, my lovely wife Corinna emailed me the latest issue of the Bird Guides newsletter, which includes an interesting news item about *Euplagia quadpunctaria.* Apparently a survey of Lepidoptera this summer has shown that this species is rapidly increasing its range along the south coast and north into the London suburbs. This proves that *E. quadpunctaria* is liable to suddenly increase its range. If it can increase its range, presumably through a population explosion, as far as north as Forest Hill and Kensington in London, and all along the south coast to the Isle of Wight and Sussex, then the possibility of them having been blown across the Channel from Jersey cannot be discounted, and indeed seems ever the more likely. Therefore, citing

the presumption of innocence which according the late Viscount Sankey is the golden thread which runs throughout the web of English criminal law, the much maligned Reverend Morris does seem, after a century and a half, does seem to be off the hook.

So when one takes on board the fact that – within the natural world as within life itself – things are never always what they seem, one finds oneself having to adopt an increasingly Fortean attitude to even the most seemingly mundane matters of biogeography. In *Wild Talents*, Charles Fort wrote: *"Not a bottle of catsup can fall from a tenement-house fire-escape in Harlem, without being noted — not only by the indignant people downstairs, but — even though infinitesimally — universally — maybe — Affecting the price of pajamas, in Jersey City: the temper of somebody's mother-in-law, in Greenland; the demand, in China, for rhinoceros horns."* This is basically the same concept as 'the butterfly effect' postulated by Edward Norton Lorenz, one of the fathers of chaos mathematics.

When I returned to North Devon in 2005, after an absence of a quarter of a century, I threw myself once again into the study of the local natural history. One of the things that I found particularly interesting was that some of the butterfly and moth species that had once been common in the region were wholly or partly vanished, whereas others which I had never seen as a child were now quite common.

One of the most important of these was the group Sphingidae – the hawkmoths. I have been interested in hawkmoths for over 40 years, ever since my Uncle Tim gave me a copy of the *Observer's Book of Larger British Moths* for my birthday in 1967.

I soon fell in love with these magnificent insects and kept and bred several species including the awesome death's-head hawkmoth, although I kept *Acherontia styx,* the Chinese species, rather than the even more striking European *A. atrophos.*

When I came to England in 1971 I looked forward to studying the British hawkmoths in depth, and was very disappointed that during the ten years or so that I lived in the tiny North Devon village of Woolsery I encountered very few of them. In 1975 someone brought me a battered and very dead privet hawkmoth which had been found in the telephone box in the middle of the village, in the same year I found two well-grown caterpillars of the large elephant hawkmoth on a privet bush just up the road, and a year or two earlier a local farmer gave me the chrysalis of what was probably a death's-head which had been found in one of his potato fields. I singularly failed to raise either the chrysalis or the caterpillars and that was the sum total of my experience with hawkmoths in Woolsery until my return.

In the last five years, I have encountered four different species in my garden alone and this reawakened my interest in the Sphingidae. There are nineteen hawkmoths on the British list, although two of these are only known from one or two specimens, and several others are very rare vagrants which do not arrive on our shores every year and have only seldom bred.

One of the most interesting hawkmoths in Britain is the pine hawkmoth (*Hyloicus pinastri*). Like the Jersey tiger moth that we discussed above there is a largely forgotten but very real shadow hanging over the legitimacy of this species. It was not even mentioned in any books on British Lepidoptera until 1800 when Donovan writing in *A Natural History of British Insect* noted that this species had been occasionally found in Scotland.

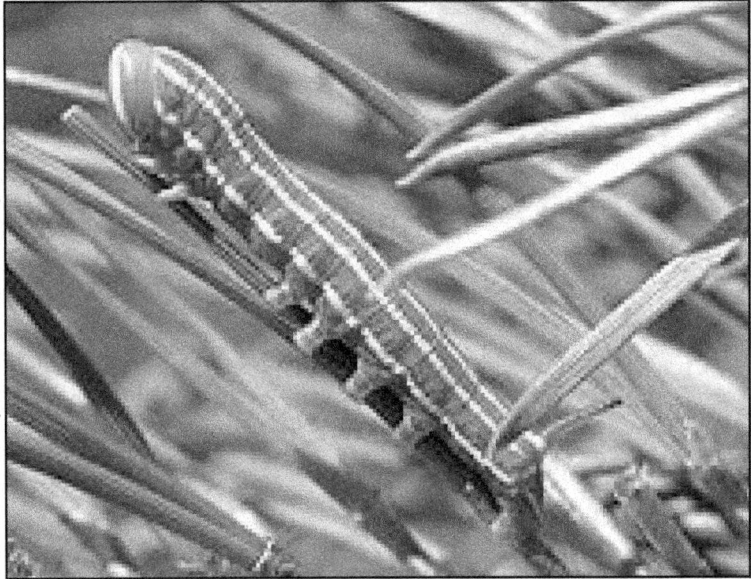

This moth has a very wide distribution across Europe, in all areas where the Scots Pine (*Pinus silvestris*) is to be found. It is common in Germany, Poland, Northern Russia, Scandinavia and many other parts of the continent. However, it wasn't until the 1880s that more and more reports became to come in from East Anglia, and – as we have seen – the entomological establishment of the time was highly skeptical of new arrivals on the British list. One suspects that another vicious turf war went on behind the scenes but for some reason the matter escaped Philip Allan who had just as much of an interest in gossip as he had in moths, so we are left with the far more sober writings of L. Hugh Newman who, in his 1965 book *Hawkmoths of Great Britain and Europe* quoted J.W. Tutt who wrote in 1886:

*"We are no nearer to any exact knowledge of the date when* S. pinastri *first came to Britain, whether before that time that the North Sea separated us from the continent (a few thousand years ago), whether with the first artificial introduction of its food plant into Suffolk (probably several hundred years ago), or with a later importation of firs, or by means of a more recent immigration. All these are things which resolve themselves into guesswork, but have no scientific value whatever."*

Newman notes that the larvae of Suffolk stock tend to be more somber in colouration than the progeny of pupae imported from Germany which would seem to indicate that the UK moths are a somewhat local form which may have been insolated for some considerable time. In the 1930s and 1940s this species spread dramatically and was found as far west as Dorset, but as of 2010 there are only a handful of Devon and Somerset records, one in Wales and a few in the north-west. The heartland of the species is still East Anglia, the Home Counties and the south-east, although it is found as far north as Humberside.

The question of its origin still remains. We have the historical evidence that it – like the Jersey tiger – is a species which can on occasion dramatically increase its numbers and range, but it does not seem to have done so in the last 70 years and so we are no nearer to discovering the mystery of its origin in this country than was Tutt 150 years ago.

Bernard Heuvelmans himself, writing in the first volume of *Cryptozoology* - the journal of the International Society of Cryptozoology - wrote that cryptozoology is the study of unexpected animals. Taking this as my dictate I feel that although it may be many years before we discover the truth behind such cryptozoological *cause célèbre* as the yeti, bigfoot and the thylacine, and whereas other icons of cryptozoology have been disproven to the satisfaction of everyone but the most gullible fundamentalist or the most cynical scion of the popular press, the study of 'unexpected animals' on a local or national level poses just as many riddles and just as many seemingly insoluble problems.

# SURINAME TOADS

**Richard Freeman**

Suriname toads are unusual amphibians that make odd and rewarding pets. They are best known for their dorso-ventrally flattened body which makes them look as if they have been run over by a steam roller in a cartoon! This is an adaptation to hiding in leaf litter at the bottoms of pools and streams. The flattened shape and grey/brown colouring provide perfect camouflage as they rather resemble leaves. A similar design is seen in the mata-mata turtle of tropical South America. In fact the faces of the Suriname toad and the mata-mata closely resemble each other in shape. The camouflage helps hide them from both predators and potential prey, but both species are also suction feeders. When they spot potential prey within range, they open their mouths and expand their throats to create a vacuum into which water flows rapidly, carrying the prey in with it. The wide, but flat heads of both species allow the mouth to be opened wide, but also for a degree of dorso-ventral expansion to create the vacuum.

There are seven known species in the family Pipidae distributed in northern South America, but the family also includes the well-known *Xenopus* species, the African clawed frogs. There are seven species in the genus: *Pipa arrabali* (Arrabal's Suriname toad), *P. aspera* (Albina Suriname toad), *P. carvalhoi* (Carvalho's Suriname toad), *P. myersi* (Myers' Suriname toad), *P. parva* (Sabana Suriname toad), *P. pipa* (common Suriname toad) and *P.*

*snethlageae* (Utinga Suriname toad). All prefer oxygen deficient, cloudy water and breathe by surfacing for air. They can reach 8 inches in length; though typically stay around 5-6 with females being larger.

The fingers are unwebbed and end in star like appendages that are used for feeling for prey in murky waters. Their eyes are vestigial and almost useless as they have little need of them in the murky waters they inhabit. The head is triangular giving the toad a truly strange countenance. Due to the strange method of feeding, they have neither teeth nor tongue because neither are needed to capture the prey, but the mouth is wide with sharp bony edges not unlike a horned toad (*Ceratophrys*).

During the dry season Suriname toads bury themselves in deep mud.

For housing they will need a vivarium several times their own length. It needs to be deep as it has to hold water. Suriname toads are almost wholly aquatic and very rarely venture onto land. Fill the tank up to about half level to simulate the shallow water in which they generally live. Water temperature needs to be 72-78°F by day, dropping to 65-74 at night. Suriname toads are best kept in filtered water. They appreciate water plants to hide in and floating wood. Be careful with substrate; Suriname toads tend to swallow food items whole in one gulp and may ingest small pieces of substrate. Use large marbles or gravel that is too big to be swallowed, the toads should spit it back out again.

Suriname toads will relish earthworms, grubs and small fish. Keeping and breeding your own feeder food allows you to directly control the health of the feeders, as well as being able to gut load them. In the long run, breeding your own live food is also often cheaper than buying it in tubs.

They can be induced to breed by raising the water level and thus cooling the water. Males do not croak, but instead produce a sharp clicking sound by snapping the hyoid bone in their throat to attract mates. The male will grasp the female behind the hind legs in a hold known as amplexus. Together they will flip through the water in a series of arcs. During these flips the female releases small clusters of eggs, which adhere to her back due to the movements of the male as he releases sperm. Over the course of several days the eggs become imbedded in pockets of skin on the females back. This is an adaptation to prevent predation during the development of the young. It appears to cause the female no discomfort. The young develop through the whole tadpole stage inside the female's back and emerge as tiny versions of the adults, each about an inch long. Incubation can take from 12 to 20 weeks and is slightly temperature dependent. Fewer eggs are produced than most other species on account of this efficient way of ensuring survival. The young need to be collected and kept separately in smaller tanks once released as the adults may eat them and the young may eat each other.

In short then, Suriname toads are wonderful looking amphibians that, providing their care needs are met, make for long-lived and unique aquatic pets.

# TETRAPODZOOLOGY
## bookone

# DARREN NAISH

Available now £12.50/$US25

# TOXIC CATERPILLARS

## Nick Wadham

*"Once a lonely caterpillar sat and cried, to a sympathetic beetle by his side..."*

If you, like me, know this wonderful song by Burl Ives, you will no doubt - by now - have the equally wonderful tune running around in your head. But for how long can something so lovely stay lovely, before it begins to irritate - before it begins to burn into your sanity? I know you've been there, with that once favoured ditty running on the endless reel of your subconscious. Before long it begins to burn and itch and eventually becomes completely so all-consuming you have to, somehow, rip open your cranium and scratch the intrusive phantom invader from your rippled grey matter!

If you ponder - in any great detail - the semantics of itch and irritate, or scratch and burn, you'll no doubt eventually (if you are an avid entomologist like myself), find yourself thinking about the amazing defence mechanisms of caterpillars. Be not mistaken; whilst a great many of them are incredibly beautiful, some of them are by the same token, just as dangerous!

You will likely now be reading this and thinking "thank goodness these are all tropical species", but woe betide your rash conclusion. Oh yes, the tropics are not the only hosts to urticating Lepidoptera, let me assure you!

In my experience, two of the most irritating moth larvae we have are those of the oak eggar, *Lasiocampa quercus*, (below right) and the fox moth, *Macrothylacia rubi*. (below left) I would argue that the fox moth is the worst, with hairs of sumptuous black velvet that belie their potent nature, punctuated by ambient vermilion rings amid an auburn helping of hair to advertise their apparent danger. Depending on the sensitivity of your skin, just a few of these hairs can create burning, itching sensations that tingle, but when the urge to scratch is too much

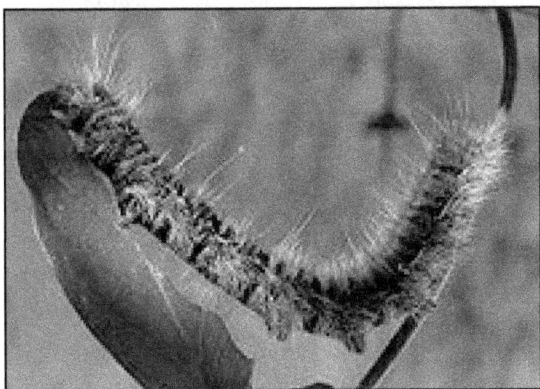

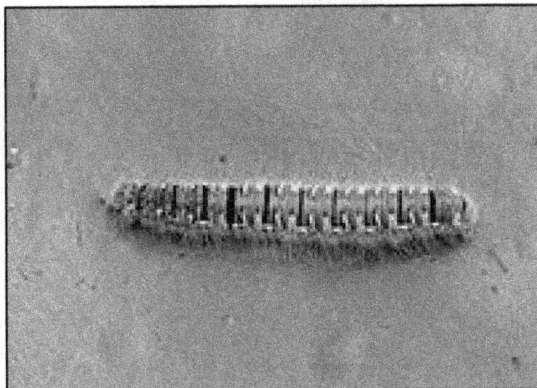

and you relent, the itch is replaced by a savage scorching sting, and your skin will be speckled by large hard white welts to remind you of the folly of your careless caterpillar catching ways!

The true domineers of defence are some of the silkmoths from the tropics. The larger saturnids, many of whom have developed fiercely stinging spines, are notoriously nasty. A recent acquisition of mine is an Emperor moth-sized caterpillar from Ecuador, *Automeris tridens*.

Though not necessarily sharp, their spines can puncture the softer skin of your palm, but if even lightly brushed against softer skin, for example the back of the hand, the effect is instantly apparent, delivering a nettle-like sting resulting in rapid localised swelling and tingling that lasts for up to half an hour, as well as white wheals.

This family is a large one at that, with many other representatives, each of which possesses similar defences - sometimes stronger!

However, there are worse caterpillars to be found!

None apparently more so than those of the caterpillars of the flannel moths! Also known as puss moths, these insects can be found in America, Brazil, and Africa. Though they may look cute and fluffy, hidden in the fur lies an armoury of sharp, brittle, hollow spines filled with an intense urticating fluid. Considering the small size of these caterpillars, (1-1.5 inches), they have a sting (varying from species to species), so powerful and painful it can knock those of a sensitive disposition unconscious. *Megalopyge opercularis* the southern flannel moth caterpillar is pictured below.

They also produce a whole range of physiological responses such as swelling, subcutaneous bruising, swollen lymph glands and nausea.

However, as Macbeth was dwarfed by King Duncan's crown and borrowed robes, neither can either of these so far mentioned larvae claim the crown of the most dangerous!

The title belongs to a rather drab and unassuming caterpillar of another saturnid known as *Lonomia obliqua*. There are a number of species of *Lonomia*, all of which are medically significant, but none more so than *L. obliqua*. "But why?" I hear you ask.

As caterpillars go, it doesn't look all that spectacular, and if you examine the safe-to-handle moth which resembles a leaf, you'd be hard pressed to think of what sets it apart in the world of Lepidoptera.

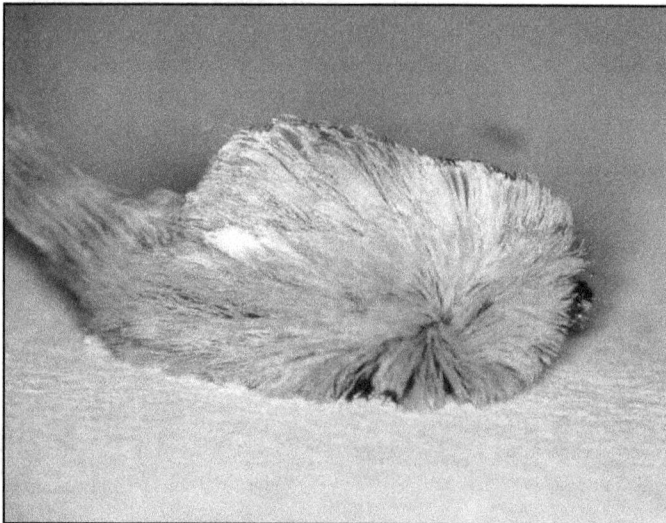

Frighteningly though, the truth is that if you so much as brush against just one caterpillar, you get stung. The pain varies and is dependent upon the size of the caterpillar and your sensitivity. If after two days of the incident you haven't either admitted yourself to hospital, or been admitted by increasingly worried friends or relatives, you will be dead!

This caterpillar administers a prothrombin inhibitor via its sharp fragile spines, and one tiny little drop is all it takes.

So far the LD50 has not yet been ascertained (*Wikipedia*). LD50 means the lethal dose required to kill half of any given test population (generally rats are used) and is worked out (in simple terms) by calculating the mass of toxin in grams per kilogram to kill the test subject.

As a measure of lethality, the death rate of encounters with *Lonomia* is around 1.7%. This may not sound impressive on its own, but let's compare it to that of the rattlesnake, whose fatality rate is in the region of 1.8% (*Wikipedia*); between 1989 and 2005 it is thought that 354 people have died as a result of the sting from these caterpillars. Shocked?

Too right! Especially when you consider that the caterpillar only administers a tiny pin prick of venom into you.

Often the encounter will involve contact with numerous caterpillars as they are a gregarious species, severe cases of which result in multiple haemorrhagic responses, including brain haemorrhage, vomiting and rapid death.

So, the next time you are invited to the ugly bug ball, just be sure to be careful about which bug you choose to hug!

# TURNING OVER A NEW LEAF

## Max Blake

Many aquarists who are interested in oddballs are familiar with both the Amazonian leaf fish, *Monocirrhus polyacanthus*, and the Nandus leaf fish, *Nandus nandus*. Both species tend to be ignored by aquarists due to their highly extendable mouths which allow them to take surprisingly large prey, thus forcing them to be kept with larger fish. This brings its own set of problems, and shows both species to be typical predators; shy to a fault. Species aquaria are generally employed, but the reliance on live food, particularly for the Amazonian species, expensive prices and the need for very clean water renders them generally unsuited for home aquaria. However, in the group (families Nandidae and Polycentridae), though there are around seven species which look and behave like the aforementioned predators, there is a small sub-family of real oddballs, the Pristolepidinae.

Rather than being obligate carnivores, these strange fish are mostly vegetarian, though they will take insects and very small fish if they can. The one genus, *Pristolepis*, contains three species, all from South East Asia down to Indone-

sia. These are all pretty similar ecologically, and only really vary in size, location and colour. *P. marginata* is the smallest at 15cm with a uniform grey-brown colouration with pale edges to the dorsal and anal fins. It has the most restricted range in the group, being found only in the extreme South West of India. Photographs of this group are unlikely to really show the species that they say they do, but as they all look broadly similar, this is not too much of a problem. *P. grooti* is the next step up in size, being around 18cm in length. It is the only species mentioned in the "Baensch Aquarium Atlas", but this gives a false impression that you may be able to find useful information about it on the internet. I have no idea what the adults look like, but the juveniles are striped in chocolate brown over a lighter brown background. However, this will probably give way to a uniformly brown colour as it grows. Finally, we come to the largest species, *P. fasciata*. At 22cm in length, it is the largest of all the leaf fish, despite not looking like one. Both it and *P. grooti* have large ranges being found throughout South eastern Asia. These turn a rich brown as they get larger from striped juveniles, and my sub-adult has beautiful blue markings just behind its eyes. There are photographs and drawings of this species on the internet, and many individuals show banding as adults. However, mine shows the banding only when stressed, and indeed images of well banded fish show them with a hook in their mouth...

My experience with this group comes from my *Pristolepis fasciata*, generally known as the tiger leaf fish. From what information I can find, they are all fairly similar and can be kept in the aquarium in the same way. They are tropical fish, enjoying warm water between around 23-28°C. They are not picky about water; I have kept mine in pH's varying from 6.5 to 7.8 and it seemed to be happy within this range. They are not fussy about general hardness either and are likely to be happy between 5-25°. They echo medium sized cichlids in much of their behaviour, being mildly territorial but not that aggressive. They seem to enjoy

caves, and some rockwork piled up at one side of the aquarium (which should be a minimum of three feet long) works well for them. Substrate is not really a big issue, so it is all down to personal preference. They may root through it a little though, so are not superb additions to intricately planted tanks. I have never seen mine really attack other fish; it prefers to mock charge some of the intruders, but leaves others. Mine spent a long time happily sharing its territory with a Siamese bumblebee catfish, *Pseudomystus siamensis,* and they nestled together in the cave.

Feeding doesn't present much of a problem. They will eat some soft leaved plants like *Elodea*, but tough or unpleasant-tasting plants like java fern will probably be left alone. Having said that, they will not rip plants to shreds, but merely pick off a leaf or two from time to time. They have small mouths, not just compared to the other leaf fish, but for fish in general. Combined with their deep chunky body, it does make them look slightly silly. Soft algae is grazed upon, but they will pick at blanket weed too. Pellets are taken with gusto, and small additions of bloodworm or beef heart will be taken too, however this should not be fed more than weekly.

Tankmates can include pretty much any medium sized community fish. Barbs (arulius and melon especially) work very well, as do larger gouramis, catfish and smaller, calm snakeheads. A riverine environment would show all these fish off to their best; flow should not be too intense however due to the leaf fish's slow and gentle swimming.

*Pristolepis fasciata* can be found in larger, more specialist stores, and smaller fish shops may be able to order them in. I have never heard of the other species turning up in the UK, but if you do see any of them, snap them up. You will be rewarded with a peaceful specimen fish ideally suited to a south eastern Asia biotope with a mixed community of medium sized fish.

# AN ATTITUDE OUT OF THE ARK

## Richard Freeman

Creationism, which in my opinion is nothing but a backward superstition, and at worst a social disease which damages society, has never held much sway in Britain. Unlike the USA, where there are many museums and other attractions peddling this nonsensical drivel as scientific fact, the UK has treated it with the scorn it deserves, until now. Genesis Expo, a creationist museum, has opened in Portsmouth, and worse still, there is a creationist zoo!

Noah's Ark Zoo Farm is a creationist zoo located near Wraxall just outside of Bristol. As a former zookeeper myself and an ardent supporter of responsible zoos, I find it horrific that an animal collection is being used to poison the minds of the visiting public, especially children. In a time were an understanding of the natural world is more important than ever before, we have an organisation purposefully and wilfully giving out misinformation.

It seems that not only is the zoo telling lies to the public, its animal care leaves a lot to be desired as well. In December of 2009 Noah's Ark Zoo Farm was expelled from BIAZA (British and Irish Association of Zoos and Aquariums), the zoological garden's regulating body, for bringing zoos into disrepute by their involvement with The Great British Circus and their failure to provide requested information. BIAZA said... *'The reasons for termination are due to a refusal to provide BIAZA with information when requested and entering into an arrangement with the Great British Circus, which contravenes the Animal Transaction Policy, despite having been warned of possible* consequences' and that *'the behavior of NAZF has brought the association into disrepute'*

The BBC and the Captive Animals Protection Society charged Noah's Ark Zoo Farm with keeping the fact that many of its animals such as tigers and camels were in fact owned by The Great British Circus from both the visiting public and BIAZA. A BBC programme showed a tiger's head in a freezer at the farm, and zoo staff were secretly filmed admitting that the tiger's carcass had been illegally buried on the farm's land, contrary to DEFRA (The Department for Environment, Food and Rural Affairs) regulations, while the head, paws and skin were kept with the intention of displaying them. The zoo's owners claimed that it did not hold a circus tiger and that their tigers belonged to Linctrek Ltd, a company that provides animals for film and television. However one of the zoo's directors, Martin Lacey, is also the owner of The Great British Circus - surprise, surprise!

Zoo inspectors also found failures to comply with the Secretary of State's Standards of Modern Zoo Practice. It was also claimed that the zoo culled animals in winter to save on their bills. Zoo owner Anthony Bush believes that the Earth is only 100,000 years old. One might as well still believe in a flat Earth around which the sun revolves.

A zoo statement runs *'To follow Darwinism is to recognize only the fleshly* [sic] *side of our natures, and, as we know, the flesh perishes; Darwinism, in other words, is a philosophy of death'.*

The British Humanist Association has urged tourist boards to stop promoting the zoo out of concern that it might undermine education and the teaching of science. Even a vicar, Michael Roberts, has stated that that the British Humanist Association was 'justified in criticising' the zoo and argued that church groups should have been more forthright in their criticism.

On Wikipedia is a photograph of a display in the zoo which purports to show the differences between apes and humans. Some of the 'scientific arguments' include that 'humans have uniquely varied beauty of face and body' that 'humans have a uniquely beautiful voice' and 'humans have a uniquely beautiful potential of inner character'. Other arguments are oddly sexual in nature, and - though true - are of little relevance to the zoo's quasi theological thesis; for example that women's breasts are attractive, men have a larger penis than apes, women have a sensitive clitoris and that hu-

mans can mate in private! At the risk of sounding like Richard Littlejohn you couldn't make this up. It shows that the directors and owners of this farcical excuse for a zoo have no grasp of science. At the risk of sounding like Richard Littlejohn you couldn't make this up. It shows that the directors and owners of this farcical excuse for a zoo have no grasp of science.

It is all the more disturbing then, that the zoo has won several national awards, including 'Silver' in the Green Tourism Business Scheme and the 'Learning Outside the Classroom' Quality Badge.

Do we really want this poison being pumped into the brains of kids? Do we want endangered animals like rhinos and tigers in the care of scientific imbeciles and superstitious morons? It's high time the animals were re-homed and Noah's Ark sunk for good.

# Revolting plants as a substitute for exotic pets

**Mark Pajak**

**www.markpajak.co.uk**

Last year a change of landlord forced me to drastically reduce my exotic animal collection. For the sake of one semantically ambiguous clause in a tenancy agreement, those creatures which could not be relocated with the mother-in-law had to go and I was left in search for a substitute. How hard could it be to find an attractive alternative to such segmented housemates? Luckily, plants do not seem to attract the same stigma as exotic animals when it comes to finding a flat and so I decided to try my hand at growing some of the more curious botanical beasts. Through this endeavour I have found that there really are some plants weird and wonderful enough to inspire and captivate the mind of the exotic pet keeper. Leopard print spots, sporadic dormancy, giving off heat, smelling like the back end of a farmyard and a peculiar fetish for flies might sound like descriptions of some strange animal, but these are all attributes of some very peculiar plants which I have had the pleasure, and perhaps even disgust, of cohabiting with. The plants in question all belong to the botanical family Araceae, also known as Aroids. They can grow more than an inch per day when first putting out a leaf or flower, which I would argue is more 'behaviour per distance' than some arachnids or amphibians that I have kept before. They have a global distribution and as such there are species to suit all climates – as houseplants, hardy garden plants and greenhouse exotics. Here I shall introduce some of these plants and try to convey how interesting they are to try and grow.

The feature that all Aroids have in common is their flower, which more technically is called an inflorescence, consisting of groups of small flowers enclosed in a chamber. Aroid flowers have developed various mechanisms to trap and attract insects, targeting in particular, flies. Unlike carnivorous plants this adaptation is not for eating insects, but for enticing them as pollinators. Although many plant-pollinator interactions result in a reward for both parties, the Aroids use deception to attract their victims into the bowls of the flowers where they act to transfer pollen from nearby plants of the same species. Flies are renowned for feeding and fornicating in such unsightly places as dung, rotting vegetation and carcasses, and it is these environments that Aroid flowers mimic – in smell, texture and colour! If you can appreciate the challenge of rearing such a plant to its horrific climax then the species mentioned below should provide many hours of fun, and they work fantastically as practical jokes too- innocuous looking bulbs (or more precisely, corms) make fantastic birthday presents.

**Voodoo Lilly, *Sauromatum venosum*. (Synonyms (i.e., may also be sold as): *Sauromatum guttatum, Arum cornutum, Typhonium venosum*)**

**Fig 1. - leaves of two voodoo lilies**

The natural habitat of this genus stretches from

**Fig 2. - Stem detail**

scribed as a leopard-print petal – the spathe.

**Fig 3. Flower and emerging flower**

Africa through tropical Asia and China. It has a seasonal life cycle, growing each year from a dormant corm into a single, pedately lobed leaf on a long leaf stalk. Each year the underground corm increases in size until it is large enough to flower. Dormant corms are sold in garden centers or markets for as little as 75p a go. They can be kept dry until spring when warmer temperatures will cause a large central shoot to emerge. Flowering sized tubers are about 5cm in diameter, and as such the shoot emerging will develop into one of the strangest flowers I have ever seen – over the course of a few weeks a 40cm conical, shiny, red-grey bag of tissue forms above the ground – swollen at the base and rising straight up to a point. The outer layer of this dark, sinister looking structure eventually splits down one side and peels back to reveal what could possibly be de-

The other part, the spadix, is thus revealed as a taut purple rod flailing high above the swollen base. The tissue of the spadix heats up to release a complex mixture of volatile chemicals that are responsible for a cacophony of vile smells that alight on the faintest breeze, reminiscent of farm smells, dung and 'carnivore crap'. The smell is more curious than repulsive – I would advise to grow this plant in a pot so as to be able to move it, on request, to a spot

away from those with a frail constitution. The flower will shrivel and rot after a couple of weeks, however this most certainly is not the end of the show – as soon as the voodoo lily retreats back into the ground, a new shoot emerges – this time to form a leaf, and true to form this plant has a very weird leaf!

**Fig 4: Flower detail**

The single leaf is held aloft on a 50cm long leaf stalk, attractively marked in dark green blotches. At the top, the leaf bifurcates and folds round on itself whilst extending into broad lobes giving the look of tropical, if not alien foliage. My current opinion is that voodoo lilies are pretty much indestructible if kept out of the frost – just plant each corm about 10-15 cm below the ground and water well in spring and summer. When temperatures drop in autumn leave them in a shed or away from rain

until next spring.

### Dragon Arum: *Dracunculus vulgaris*

*Dracunculus vulgaris* has a Mediterranean distribution, with various morphological forms found throughout various Greek islands (Bown, 2000). The plant has a similar growth pattern to voodoo lilies; however the flower is produced in the same growing cycle as the foliage, and several lobed leaves sprout from a single corm. If the voodoo lily smell is curious, then the stench of the dragon arum can only be described as pure evil.

Planted in close proximity to neighbours, or worse indoors, it would certainly raise suspicion that a large mammal had died somewhere and begun putrefaction. Overpowering as it is, the smell is ingenious, and one can spend hours watching all sorts of flies, beetles and strange looking wasps land on the spadix in a frenzy. Dragon arums seem to prefer very sunny spots, and I have found that plants grown in shade do not gather enough momentum to flower the same season. These can also be taken out of the ground or pot when the last of the foliage has withered in autumn – to be replanted when they begin growing again in spring.

**Fig 5: Dragon arum and flower bud**

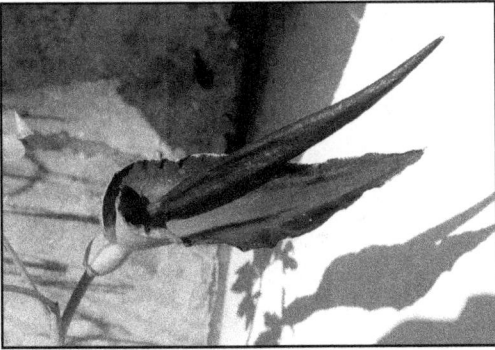

**Fig 6: Dragon arum flower**

**Fig7:  Dragon arum corm**

## Snake Palm *Amorphophallus konjac* (Synonyms *Amorphphallus rivieri*)

The genus *Amorphophallus* contains some very large and very smelly inflorescences: The titan arum or corpse flower, *Amorphophallus titanum* can grow to over 3m tall in the jungles of Sumatra. Clearly this species is not possible to cultivate in the average living room or even garden. The Botanical Gardens at Kew first raised a plant to flowering from seed in 1889 and subsequent flowerings across the world usually attract media attention, with somewhat disturbing images of a gloved and gas masked horticulturalist attempting artificial pollination through a window cut through the outer spathe tissue layer. The 200 odd species of *Amorpho-*

*phallus* are distributed across the tropics of Asia, Africa Australia and Melanesia, and several different species are easy to get hold of at a low price, although rarer species fetch large sums on Ebay. The snake palm, *Amorphophallus konjac* is one of the easiest to grow, and can be treated in exactly the same way as the voodoo lily and dragon arum. This plant will take a few growing seasons to bloom, and having started with small tubers I have not yet reared a plant to fruition – this is more of a challenge than those listed above, but the vital statistics, and for that matter the smell, will make this all the more worthwhile to attempt: The inflorescence can exceed 50 cm, held up on a leaf stalk 5 feet tall!

**Fig 8: immature snake palm**

## The Dead Horse Arum: *Helicodiceros muscivorus*

The name says it all really, and this is another plant grown for the challenge of producing a flower. It is closely related to *Dracunculus*, requiring a sunny location as per its native home of the Mediterranean islands where it must compete with the smell of death and faeces amongst seabird colonies to successfully attract pollinating flies and beetles (Bown 2000). The plant is somewhat of a challenge to grow since the bizarrely shaped leaves are a little flimsy and tend to break if unsupported. It is best to plant the tuber as deep as possible to provide maximum support for the leaf stalks. Although challenging, one of the paybacks is that the tuber will readily produce offsets, which in time form many small plants so you can try out several different growing locations to increase success. The inflorescence of the dead horse arum leaves little to the imagination – over 30cm long and broad and covered with thick hairs – mimicking the rear end of a deceased animal – the hairy spadix protrudes tail-like from a dark pit down which flies crawl and are trapped overnight in the floral chamber thus facilitating pollination. In keeping with many other Aroids, the flies receive no reward for the act of pollination, and to top it off any maggots hatching from the eggs they lay will starve.

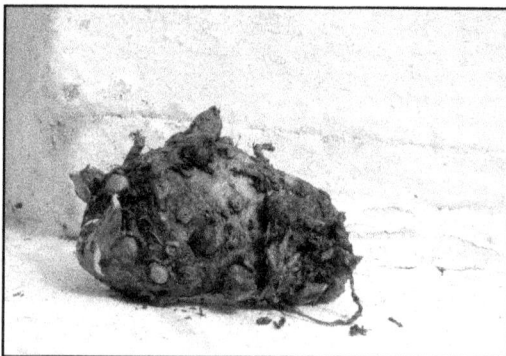

**Fig 9: corm of dead horse arum**

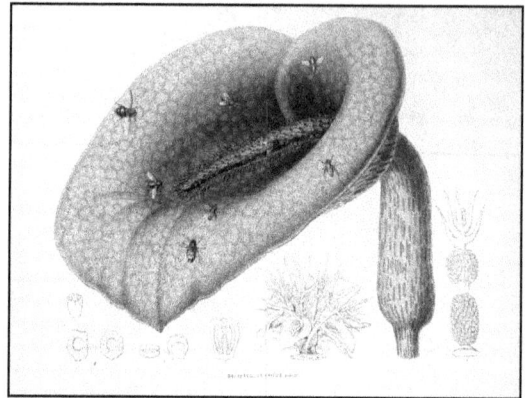

**Fig 10: illustration of the inflorescence of the Dead Horse Arum**
(from http://upload.wikimedia.org/wikipedia/commons/e/e3/Helicodiceros_muscivorus00.jpg Copyright: public domain)

## In Summary:

For the exotic animal keeper that does not need to personify their pets, for those who see beauty in strange places, for want of something weird that does not need to be fed, Aroids fit the bill a treat. They are inexpensive, for the most part easy to rear, exciting, revolting, and certainly a catalyst for conversation: I would recommend anyone to risk giving them a go. For me they have revealed many complex and wonderful interactions of the natural world, instilled a passion for our native Diptera and opened my nostrils to new experiences. If however you decide that these plants are better left in the dark damp tropics from whence they came then I have some unfortunate news for you – Lords and Ladies, *Arum maculatum* is a common woodland species and a fantastic Aroid with a fiendishly cow-pat-esque stench.

**Further information:**
The International Aroid Society www.aroid.org
**References:**
Bown, D. 2000. Aroids. Plants of the *Arum* Family (2nd ed.). Timber Press
**Online retailers:**
Ebay – search for Aroid, Aracea, Arum, Amorphophallus etc.
www.himalayangardens.com

This is a photograph of the flower of *Amorphophallus titanium*, the titan arum. A close relative of the other Araceae covered in this article, titan arums have the largest single unbranched inflorescence in the world. Interestingly, and a little disturbingly, the scientific name translates to "giant, misshapen penis". The parasitic *Rafflesia arnoldii* has the largest single flower with a diameter of three feet (titan arums have a group of flowers technically called an inflorescence), whilst *Corypha umbraculifera*, the talipot palm, has the largest branched inflorescence at between six to eight meters. This photograph shows Dr Chris Clark and John Hare with one of these flowers in Sumatra in 2004 during a CFZ expedition to find more evidence to support the existence of Orang-pendek, an upright walking ape likely related to Orang-utans.

# THE SAIL-FINNED WATER DRAGON

**Richard Freeman**

The sail-finned water dragon, aka the sailfin dragon or the sailfin lizard, vies with the Australian frilled lizard for the title of the world's largest agamid. Males may reach an impressive four feet in length whilst females stay about a foot shorter. They support a dorsal crest and a spectacular fin on the tail much like those of basilisks, but it is somewhat larger.

There are 3 species in the genus *Hydrosaurus*. *H. pustulatus, the* Philippine species, can be grey, brown or green with spectacular violet or blue markings in adults. These are particularly prominent in the males. *H. amboinensis* hails from Indonesia and Papua New Guinea and generally has yellowish or dirty orange patternation against a dark grey background. *H. weberi* of the Moluccan Islands has a finer speckling of yellow.

These are semi aquatic lizards that can swim exceptionally well. They are also adept climbers often found in gallery forests. The Philippine species is now classed as vunerable due to deforestation, and its exportation has been

banned. Ergo *H. pustulatus* is uncommon in the pet trade and is highly expensive to buy. Most wild caught ones these days are *H. amboinensis.*

Despite having sharp teeth in powerful jaws, hooked claws and a strong tail, sail-finned dragons are generally quite tractable beasts and are not usually inclined to bite. Big specimens may accidentally claw their owners if they start to flail their limbs about. They should be held by the back of the neck and from underneath by the base of the tail to support the animal. They quickly become used to being handled.

These lizards have quite a long lifespan, up to 25 years in captivity.

Sailfinned water dragons are omnivores. In the wild they eat a variety of fruits and leaves as well as invertebrates and small vertebrates. The captive diet should consist of 50% plant matter and 50% animal protein. Plant food can consist of strawberries, blueberries, raspberries, melon, mango, papaya, grapes, apple, pear, banana, cucumber, shredded kale and grated carrot.

Animal food can consist of lean meat, freshwater fish and crustaceans. Sailfins will also take defrosted mouse pinkies. They are avid eaters of insects such as crickets, king mealworms and locusts. Gut loading the insect prey is a good idea. I find that fish flake food is eaten by both mealworms and crickets and is full of vitamins. Obviously smaller specimens need to have more finely chopped plant material and meat/fish as well as smaller insect prey. Feed adults around three times per week. Plant and meat/fish needs to be offered on a dish. Insects can be released into the enclosure so the lizards can hunt them as they would in the wild.

Adults are big lizards and require a large vivarium. A single adult will require a tank at the very least five feet wide by eight feet long and five feet tall. Converted garages and conservatories may make for good homes as long as they are correctly lit and heated. In general, specially made enclosures are better for sail-

finned dragons than the commercially available ones. They require access to clean water for drinking and bathing. Sailfins appreciate larger enclosures with water containers big enough for them to swim, or even built-in pools. They also like thick, sturdy branches on which to climb. If your enclosure consists of glass panels, cover three sides of the vivarium with backdrops or non-toxic paint. These are forest creatures and having light and onlookers from all sides can make them jumpy.

The substrate should be able to retain some moisture. Additive-free potting soil is fine, as is a mulch of cypress bark. This should be about three inches thick. As rainforest animals, the dragons need moisture in the air. Their humidity should be kept at around 75%. This requires them to be sprayed or 'misted' with water several times a day. Spray both the enclosure and the lizards themselves with luke warm water. If the humidity falls too low then the lizard's skin will become too dry and its circulation will be effected. This can lead to a loss of toes when the animal sheds its skin. Lack of moisture can also cause kidney disease. A container full of sphagnum moss helps keep humidity up. Thermometers and humidity indicators are a must!

Sailfins require a temperature of 85-90 degrees Fahrenheit during the day, dropping to 75-80 at night. Temperature can be controlled with a thermostat. Heat sources can be under floor heat mats or spot lamps. If you are using a ceramic heat bulb, be sure it is securely caged so the animal cannot burn itself. A basking area with a higher ambient temperature of around 115 degrees Fahrenheit can be achieved with a spot lamp position onto a branch.

A UV light (mercury vapour or fluorescent) is required to allow the lizards to process vitamin D. Make sure the light source is soundly attached as these big strong lizards can accidentally pull UV lights out of their sockets. A photo period of 12 hours dark and 12 hours light works best.

Sailfins are messy and will require daily water

changes and a daily gathering of spilt food. Males are bigger and bulkier than females with larger tail fins and more prominent nasal knobs. Males also sport larger pre-anal and femoral pores. In comparison with many other popular lizard species, sailfins are not often bred in captivity by pet owners. With some wild populations dwindling, captive breeding is highly important. Dragons will need to be at least two years old before they show interest in breeding. A reduction of the light period by about two hours and a lowering by several degrees of the heat should trigger mating behaviour.

Males will display with head bobbing and mount the female, biting the back of her neck. During courtship and mating these lizards make a curious call akin to a baby crying! After mating, a deep container full of substrate such as potting soil or vermiculite needs to be added to the enclosure. The female will lay 2-8 eggs that need to be incubated at 82-85 degrees Fahrenheit. This is best done in a separate incubator to avoid damage.

Young will hatch after two months and should be kept in their own enclosure which can be set up like a miniature version of the adults' vivarium. Their diet is broadly the same, but with much more finely chopped food material (in fact, soft fruits can be pulped) and smaller insects such as micro crickets. Young dragons tend to eat more insect food than adults and their diet should reflect this. The also like branches and hollow logs to hide in. Remember as with the adults, keep them moist. Growing young will need feeding much more than adults.

All in all, sailfin dragons take a lot of time, effort and money, but they are one of the most attractive and impressive of all lizard species and are certainly worth the expense!

# BIRTH, SEX AND DEATH IN RURAL ENGLAND

## Carl Portman

I spend a lot of time thinking about wild exotic foreign places, crying out to be explored with new species waiting to be discovered - faraway lands with exotic spiders the size of a Frisbee, colourful insects larger than life itself and all this accompanied by the steady hum of cicadas. Of such things my dreams are made. However I understand that it is important never to dismiss or study what I am lucky enough to have right here at home – right under my nose. I decided to reflect on the months just passed and revel in the myriad creatures that have visited Orchard House. Here, in Banbury there have been some exciting discoveries. Make some tea, cut yourself an unfeasibly large portion of fruit cake and let me take you on a little natural history journey around my home and garden. We'll begin indoors, appropriately with glorious new life, but end my journey later in the article with sudden and shocking death. The new life came in the form of spiderlings that were the progeny of a large female *Pholcus phalangioides* spider. Like tiny leggy dots in the constellation of my ceiling they gathered around the mother prior to dispersal around the house. There are a large number around the homestead now and the significant reduction in the number of mosquitoes and flies this year is certainly no coincidence.

Here's the spider and her young in all their glory. They appear to be tremendous mothers, carrying the eggs in their mouths for many days. I have never seen an adult – no matter how thin and hungry they must be – eating any of the young.

I observed one of the spiderlings catch its first mosquito – a most interesting moment. The initial tentative steps were suddenly superseded by a desperate huge lunge into the unknown; and instinct claimed its very first meal.

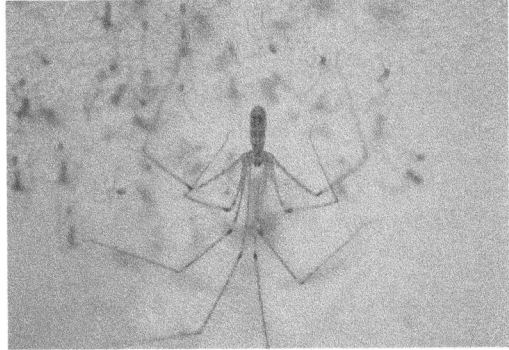

This delightful spider has a lovely defensive mechanism when disturbed. It shakes violently spinning around in a circle, perhaps imitating a leaf in the breeze so that any predator is fooled into thinking it is not a prey item. How very cool. I have seen these thin spindly arachnids catch and devour much larger *Tegenaria* spiders, and indeed, urban legend has it that *P. phalangioides* are the most venomous spiders in the world. However, because their fangs are unable to penetrate human skin no one seems to be particularly bothered. Studies have concluded that the venom is only mild so there's no need for widespread panic across the nation.

It is known as the daddy-long-legs spider amongst other things but that's confusing because at least two other invertebrates are called that. I just call it *'Pholcus'* and enjoy having them around the house.

I was delighted this year to see my old friend *Melolontha melolontha,* otherwise known as the cockchafer or May bug. These seem to be greatly feared amongst my friends and acquaintances but this unfortunately is down to ignorance. They are harmless and very beautiful in design. The size frightens people; they are after all substantial insects.

I found this one on the kitchen ledge one morning and was only too happy to take it outside and set it down on the fence near the honeysuckle. I became curious about its lifestyle and was fascinated to read that it feeds on the leaves of trees and shrubs, but that the fat grubs feed on the roots of many shrubs, including crops. Perhaps it is not the farmers' favourite then. I vividly recall seeing the aftermath of a swarm of these when I lived in Germany. It was the morning after the night before and there were thousands dead and dying by woodland on the floor of a local car park. It was a terrifically sad sight. I was powerless to save a single one...

Let's stay outside...

One morning whilst taking a leisurely breakfast outside in the warm sunshine I observed several frantic hoverflies on a dandelion. I couldn't resist the urge to find my camera and get up close and personal with them. How polite they all were; taking it in turns to feed upon the pollen whilst giving each other plenty of space. This little jewel is commonly known as the marmalade hoverfly *Episyrphus balteatus* and is common in Britain from March to November. It has been known to migrate in swarms from continental Europe.

What wonderful masters of flight they are. Sitting outside, they would often alight on my brightly coloured tee-shirt or hover right by my lips as I sipped tea. They are just fantastic and I never tire of seeing them. I have spent hours studying them but I am still light years away from having a proper understanding of their

lifestyle. If you hold your arm out and extend your finger they will often zoom in and then land on the digit for a rest. It's very rewarding to get so close to nature.

From air to the floor - I had a visit from the amphibious fraternity this year. Even though I live in small village with a small stream running past my house I don't see 'common' frogs that often. They don't seem to be that common. I do see toads aplenty. I was therefore delighted to find this rather bewildered looking *Rana temporaria* hopping along in my back garden.

It's fairly easy to tell if it's a common frog or not by the dark 'mask' on its face. Don't go just by the colour either because they vary a lot between green/olive and brown. On reflection I am sure I have heard the croaking noises from these animals when I am in the garden and I shall try my best to make my small pond more attractive to them. Unfortunately this pond is out the front so it won't be easy!

I have not actually seen frog spawn since I was a kid – that's very sad. Either I am not getting out enough or I am going to the wrong places.

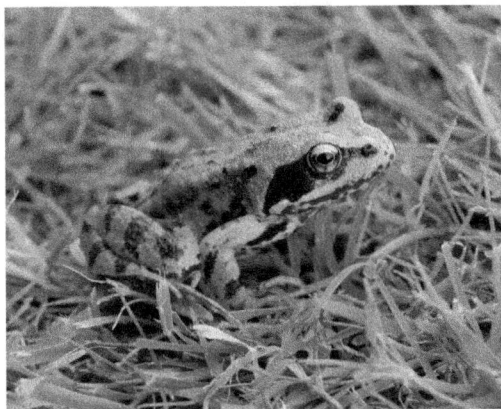

Now for a new discovery – for me anyway! I did find one insect I have never seen before. Indeed there were two specimens landing on my devil's teardrop plant. I photographed them (eventually) and ascertained that they were sawflies of the species *Tenthredo scrophulariae*. These are not very different from *Tenthredo arcuata* which has all black antennae as opposed to the gorgeous bright orange owned by *T. scrophulariae*. Apparently the larvae feed on mulleins and figworts. The adults are carnivores, feeding on flies and any other small insect it can overpower. It's another wasp mimic by the look of it. Figworts themselves were used as herbal medicines for skin eruptions in days gone by.

Now for the sex. Don't get too excited! I poked my nose into the very private life of these weevils. I think they are vine weevils *Otiorhynchus sulcatus*. I was very pleased to "catch them at it", although I should not be pleased that they were here since they are a serious destroyer of plants, eating the roots and young shoots.

This includes my nice potted plants!

Oh well, it *was* summer and even the insects want to do what comes naturally 'Alfresco'. Can an insect look embarrassed? The one on top seems to be. Still, I expect there has been a fly on the wall of many encounters between humans, but that's for another natural history readership, not here!

My penultimate piece is about another discovery at my house – this time of the mammalian variety. We have a bat roost!

How wonderful is that? After Susan's initial shock we both set about reporting, identifying and recording the bats. The loft is now a protected area and we hope to have these gorgeous creatures honour us with their presence for many years to come.I thought they were noctule bats *Nyctalus noctula* but they usually roost in trees. They appear to be common pipistrelles, *Pipistrellus pipistrellus*.

The problem is, by the time the official 'bat man' visited to record and identify them they had all flown the roost. The droppings were examined and it is likely that they are indeed Britain's smallest bat.

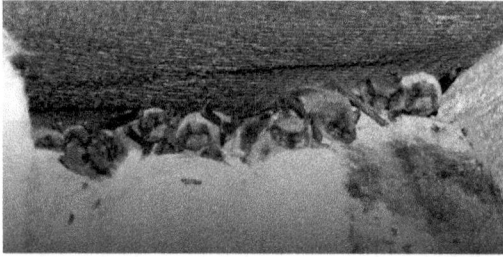

I started to investigate and learn more about bats and was shocked to find that these tiny mammals can live for up to 16 years and consume 3000 insects in a single night. Another reason why we have had no trouble with mosquitoes this summer... what a symbiotic relationship with wild creatures we have. We give them a free house in exchange for them devouring the annoying mozzies. Of course, they do eat moths too, and I hope that's not the sole reason why I have not seen so many this year, including *Macroglossum stellatarum* the hummingbird hawk moth.

The bats will return, of that I am sure. I contacted the previous occupants of the house to be told that they saw bats here in 1999.

Finally then, to death.

Spiders eat flies and almost nobody cares. But when I see a beautiful bee ensnared, my instinct is to rescue it. Rightly or wrongly I leave it, something must die so that something can live, that's nature's way 'red in tooth and claw' as the saying goes. This huge orb web spider *Araneus diadematus* quickly subdued the bee in such a way that there was never any chance of the spider being stung. Nature is in many ways a reflection of life, and we can learn much from it. The message I get from this is to know the strengths and weaknesses of your opponents. Utilise your strengths and capitalise on their weaknesses. Opportunity should never be overlooked...

I hope you can see from the prose above that there is so much to appreciate right here only metres from where you eat and sleep.

I could have mentioned the tremendous magpie moth, the snails, the variety of beetles, the scorpion fly, the shield bugs and the dragonflies – and I have not even mentioned plants! All of this I shall carefully store in my mind for another day. A day when sitting beside the fireside, old and grey and full of sleep I can reflect upon the wonder of life around us that many people simply never notice.

I must mention the wonderful line proffered by Colonel John Blashford Snell at the Weird Weekend 2010. He said that we are all driven to do what we do by curiosity. That's a small line but a very powerful statement.

It's completely true, only by being continually curious can we appreciate the things that the world around us has to offer. Only by being curious and asking questions can we hope to get the answers we crave. Nothing comes without sacrifice, in terms of time and effort but actually, much of what I have written about in this article has literally come to me. No airports, no travel sickness, no dodgy food and no delays. Now get out there and enjoy it all for yourself.

*"Whatever nature has in store for mankind, unpleasant as it may be, men must accept, for ignorance is never better than knowledge".*
Laura Fermi.

# RELEASE THE BATS

## OLL LEWIS

## Introduction

Fruit bats are quite rare animals to keep in captivity in the UK, even by exotic pet standards, but the number of people keeping them has been slowly rising here. Most people, even some regular keepers and breeders of exotic pets, are unaware that fruit bats are kept as pets, but that is changing and it is reasonable to expect that as people become more aware of them, fruit bats could even approach the popularity of pygmy hedgehogs and other exotic small mammal species.

Many different species of fruit bat are available on the market, but one of the most popular in the pet trade is the Egyptian fruit bat, *Rousettus aegyptiacus*, and this care sheet is written with that particular species in mind. However, this could also be used as a general guideline for most other Megachiropterid species provided the enclosures and dietary requirements are scaled up accordingly. Prices for *R. aegyptiacus* vary; the author has seen pairs going for between £150-£300.

It should also be noted that fruit bats can carry rabies so care must always be exercised when handling them, especially in places like the USA where, unlike the UK, rabies is prevalent. Always purchase fruit bats from a reputable breeder because there is a chance that less trustworthy breeders may try to bypass quarantine laws. This could either be to get animals on the market quickly or through not being aware that such laws apply to all mammals entering the country.

Despite media reports to the contrary, there have been no rabies infections acquired in the UK since 1902; what is usually mistaken for rabies is often European Bat Lyssavirus (EBLV) which is from the same family as rabies. In some rare cases EBLV has crossed the species barrier but if treated promptly patients can expect a full recovery. Rabies vaccination also works on EBLV and for this reason it is recommended that anyone who will be in regular contact with bats should be inoculated as a precautionary measure.

## Environment

The Egyptian fruit bat has an average wingspan of around 60cm (2ft) so any enclosure must have adequate flight space. For this reason, it is not recommended that they be kept indoors unless an entire room can be spared for them. If a room can be spared within the house then ideally it should not be carpeted and be free of fabric covered furnishings, to ensure easy, hygienic cleaning. A much better option than keeping the bats indoors is to keep them in a large aviary. The aviary must have a substantial indoor area that can be used as a roost by the bats, and heated in the winter.

The aviary must be securely put together using thick chicken wire with a mesh size of no bigger than 1cm square, not only to stop the fruit bats from escaping but also to stop - as much as is possible - other animals and birds from entering the aviary and taking the fruit bats' food. The mesh will also have to be strong to withstand possible gnawing from the bats and escape attempts. An 'air-lock' is also recommended at the entrance of the aviary to prevent escapes.

The aviary should contain a feeding table (a garden bird table will be adequate for this purpose) and several perches. Perches can be made of any kind of wood or tree branch provided they can support the animal's weight (around 160g) and be kept off the ground. Bats will prefer an aviary with plenty of foliage to hide in or shelter under, but still with adequate space for them to stretch out their wings to fly. If the bats are housed indoors rather than in an aviary then large pot plants can be used to provide adequate foliage for the bats.

In the indoor space it must be ensured that there is adequate provision of roosting spots. In the wild, fruit bats will roost high on trees where the leaves shield them from the sun's glare during the daytime or in caves, so it is best to place roosts as near to the ceiling as possible and anything that the animal can hang from comfortably and safely can be used as a roost.

Because fruit bats are nocturnal it may be preferable to include lighting in their enclosure in order to facilitate viewing when they are at their most active. Do not use bright lights as these will dazzle the bats and cause confusion to their natural sleep patterns. Always use as dim a light as you can find and place a shade over the light in order to lower the chances of the bat looking directly at the bulb. Most bat keepers report that putting a red filter over the lighting system also works well. To avoid startling a fruit bat, never point a light directly at a bat when you turn it on.

## Diet and Health

Fruit bats are, as their name suggests, frugivores. They will take most juicy or soft fruits that grow in their natural environment and these should compose the bulk of their diet. A good mixture of different fruit including apples, pears, melon, bananas, pineapples, mango, papaya, grapes and tomatoes should be fed each day. The quantity of food the fruit bats eat will vary a lot so always ensure that you have a feed surplus of fruit and be prepared for wastage as the fruit bats will rarely take more food than they want or need.

In order to ensure that all dietary requirements are met, fruit should have a very light sprinkling of vitamin and supplementary powders dusted upon it and a small amount of low fat dried cat-food can also be offered on the feeding table. Salad leaves and cress, although not lettuce, can be offered occasionally as well as the normal amount of fruit, and will sometimes be taken.

Fruit should be left mainly on the feeding table but also in other places around the enclosure and all uneaten food must be removed.

As the amount of fruit each bat will take is dependent on a number of different factors, such as how much exercise the bat has had, sex, age, if it is nursing young or how dominant it is within the group, it is difficult to state a definitive figure on how much food you should be feeding your fruit bats. This is best determined by ensuring that there is always fruit available to them and monitoring how much the fruit bats are eating for the first few weeks with you until you get a clear idea about the amount of fruit each bat will take.

To prevent boredom it is worth placing some fruit, particularly the fruit that proves most popular with the bats, in places other than the main feeding dish. These can be hung from trees and perches or partially hidden in foliage, ensuring that the bats will fly all around their environment searching for their favourite fruits rather than just flying from their roost to the feeding table and back. As well as providing good mental exercise for the bats this will also help to prevent bats from becoming over-

weight.

A good clean supply of water should be provided in large shallow bowls that cannot be tipped over. The bats will use these bowls for drinking and to bathe, so it is important that they be washed out every day and sterilised with boiling water before being refilled with fresh water.

Because properly housed captive fruit bats are likely to have little or very low contact with indigenous species of bat, the risk of the bats becoming infected with European Bat Lyssavirus (EBLV) once they are housed is very small indeed, especially because the virus has been shown to not readily transmit through oral and inhalation routes. According to research by DEFRA the virus will not pass through grooming and general handling but only through a

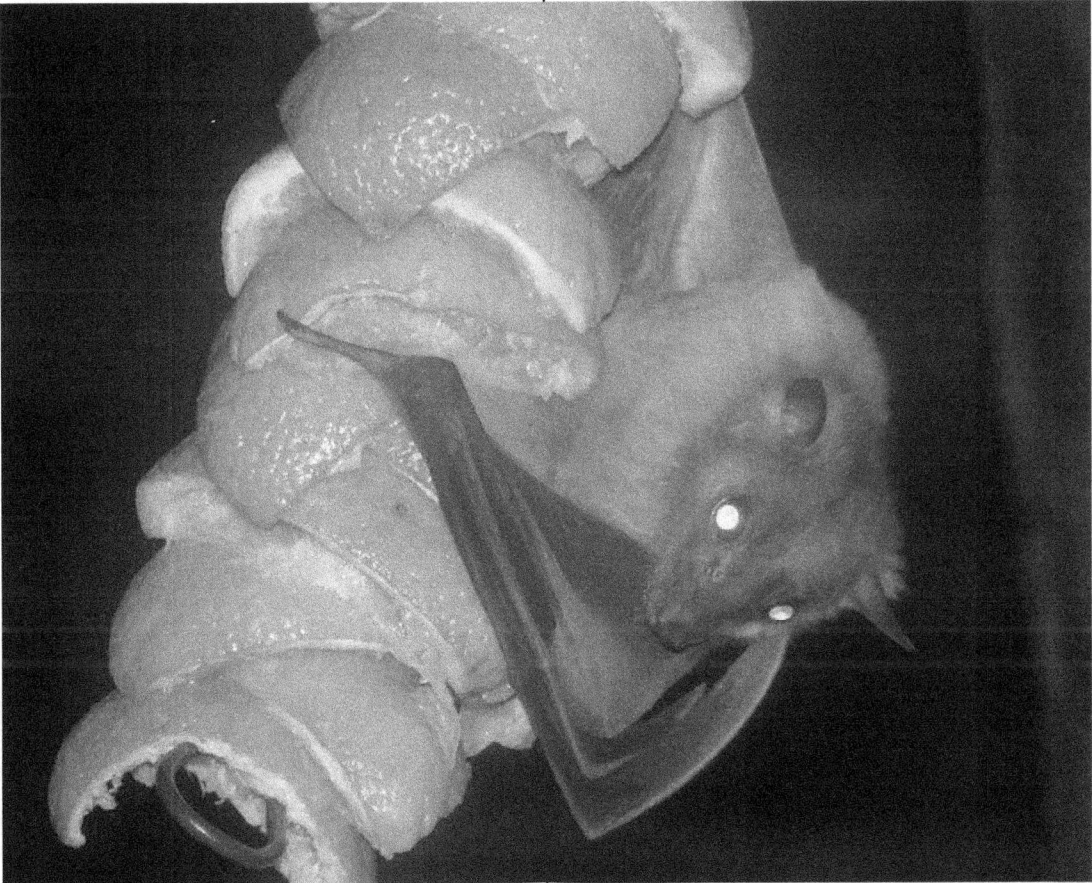

bite from one infected animal to another (including humans in some rare cases). Sadly, bats showing symptoms -which are similar to rabies - are likely to be beyond help and will likely have to be euthanised by a vet, but it is possible to save an animal if the virus is caught early. This means that all the bats not showing symptoms should be quarantined and treated by a vet as a matter of urgency. As mentioned in the introduction section, human infection is rare but if you are bitten by a bat that you even suspect may be infected with EBLV do not take any chances and seek medical treatment immediately.

## Socialising and Breeding

Fruit bats are social animals and should never be housed alone as this will often cause distress to the animal. Male fruit bats may squabble if there is a significantly larger number of males than females in the colony, so it is best to ensure that the male to female ratio is fairly equal or weighted towards females. Males are easily recognised as adults by being substantially larger than females. Fights and skirmishes between males are usually not serious and are often little more than posturing to show dominance within the group.

Socialising a fruit bat to human contact should ideally take place just after it becomes fully weaned as older fruit bats will require more effort. Socialising is achieved by holding the fruit bat or letting it roost on you or your hand for a while so that it will get used to your smell and learn that you will not hurt it. This should take place daily over several weeks and when the bat is used to human contact the amount of contact and regularity can be reduced. Bats used to human contact will readily take fruit from your hand without hesitating.

Breeding fruit bats is not difficult. Provided they are comfortable in

their enclosure and are being fed correctly the bats will breed. Female Egyptian fruit bats will typically reach sexual maturity between 15 to 16 months and males between 14 to 18 months. When the bats are ready to breed the female bats in the roost will separate off from the male bats to form a nursery roost.

Females are capable of two gestation periods a year but usually will only take advantage of one, giving birth to single or occasionally twin young after a four month gestation period. The young bats will take their first flight when they are around two months old.

## Conclusion

The Egyptian fruit bat makes an interesting and unusual pet, but should only be kept by experienced pet keepers with enough space to keep them; they are particularly unsuitable pets for people with small gardens and/or small houses. Egyptian fruit bats can live for 25 years in captivity so it really is not recommended to keep fruit bats if you are not settled. It is also a wise precaution when keeping any species of bat to protect oneself against rabies and related viruses. If you can meet all these requirements then you should be able to keep Egyptian fruit bats responsibly.

# THE SCALES HAVE FALLEN... OR SOMETHING!

## A field report from Denmark

## by Lars Thomas

Denmark has never been a country to make herpetologists foam at the mouth. Our herpetofauna is distinctly meagre – two snakes: the adder and the common grass snake; the slowworm; two lizards; and a tiny handful of frogs and toads. They are fairly unevenly distributed, as Denmark consists mainly of islands – American author Bill Bryson likened the look of Denmark to a plate dropped from a considerable height onto a hard floor – so some parts of Denmark completely lacks species that are extremely common elsewhere. Things have been rather quiet for a considerable number of years, but during the last year, everything appears to have gone berserk; maybe not exactly cryptozoologically, but distinctly weird.

There are always the escapees of course. We get our fair share of those, but this year... The odd python or garter snake gone walk-about is nothing new, but a king cobra in the central square in Copenhagen? And what about the two boys who claimed to have seen a rattlesnake on a field outside Roskilde some 30 km's west of Copenhagen? And then we have the very weird ones – like the couple going home by car from a meeting in Herning in western Denmark. It was late at night in the middle of April, and when they were getting close to their house, they suddenly saw a naked and rather curvy girl standing in the middle of the road, carrying a very large green snake draped over her shoulders. The couple stopped the car, but when they got out to investigate, the girl and the snake had both disappeared.

And even the common species are behaving strangely. All the books on Danish wildlife will tell you that the grass snake is not to be found on the large islands off the southwest coast of Denmark, but nevertheless scores of people now claim to have seen it on the island of Rømø. In places where adders have not been seen in years, they are now suddenly starting to appear in droves, and we even have people claiming sightings of 2-3 feet long slowworms! None of these have been longer than normal when investigated, but still. And the lizards – oh wow. Some have cropped up in rather special places – one on the 7th floor of a block of flats in the outskirts of Copenhagen, and some have been sporting unusual colours – blues, reds and orange, and in one case two tails.

The amphibians have been breeding like mad as well; in some cases tadpole counts have been off the charts. In tiny ponds where you will normally only find a couple of hundred or maybe a couple of thousand tadpoles the numbers have been up way beyond 50,000 and sometimes even more.

Oh yes – all the other creatures are behaving rather strangely too. The fox was exterminated on the extreme eastern Danish island of Bornholm many years ago, but just in 2009 it has been seen on Bornholm at least 15 times.

And the bugs and assorted small beasties – oh, to be an entomologist!

Something like 30 species never before seen in Denmark before have been found during the twelve months of 2009, and several rare or fairly rare species have started to crop up in the most unexpected places too. The stag beetle hasn't been seen in Denmark for something like 40 years – lo and behold, suddenly it appeared in a garden just north of Copenhagen.

The blue oil beetle, something rather special in these parts, has suddenly decided to show its fat behind far more than usual.

I found my first specimen in the spring, never having seen one in Denmark before. Central Copenhagen where I live, is not exactly a zoological haven, but nevertheless I have found 11 species of ladybird beetles in an area where one would normally only find the common 7-spot variety. I know the climate is changing, but come on!

I am sure Charles Fort is laughing himself silly in whatever heavenly libraries he hangs out these days.

# Tell Me Y

## Jonathan Downes

The Silver Y moth (*Autographa gamma*, previously known as *Plusia gamma*) is one of the most well known migratory moths on the British list. The larvae are extremely polyphagous and according to Nancy Fraser (2000), have been recorded feeding on at least 224 plant species. Furthermore, Fraser claims that they have also been recorded damaging many crops, particularly those of the cabbage family.

They have a complicated life cycle. The adults make seasonal northward migrations into areas where, due to climatic conditions, they are un-able to establish a full-term presence. As Fraser writes: 'For example, individuals migrate into Britain each spring, and after, one, two or three generations, descendants of the spring migrants return to over-wintering sites in North Africa and the Middle East'.

The species is widespread across Europe, parts of Asia and North Africa and the northerly migrations can reach as far as Iceland, Finland and Greenland – the latter, as far as I can tell, being its only known incursions into the New World. In the United Kingdom they are present in significant numbers from the middle of May until

they are killed off by frosts in the late autumn. Many individuals, however, fly south again and winter around the Mediterranean and the Black Sea.

In 2007 the Vermont Co-operative Agricultural Pest Survey (CAPS) reported that although they had not actually found any specimens "the likelihood and consequences and establishment by *A. gamma* have been evaluated in a pathway-initiated risk assessment conducted by the Department of Entomology from the University of Minnesota, published 2003.

> "*Autographa gamma* was considered highly likely at becoming established across the US if introduced: the consequences of its establishment for US agricultural and natural eco-systems were also rated high (ie., severe)."

As far as I am able to ascertain, there have been no records, however, of this species in mainland North America. That is, until (possibly) now.

As editor of an increasingly popular online daily magazine covering, amongst other things, natural history, cryptozoology and out of place animals, I often receive photographs for myself or our readers to identify. These pictures were sent on 8th August 2010 by my friend and colleague D.R. Shoop in Minnesota. They had been taken the previous Monday (the 6th). My first thought was that they were indeed a Silver Y moth.

I have been familiar with this species since I first became interested in British moths at the age of eleven, 40 years ago. My second thought was that they couldn't be because Silver Y moths are not found in North America. Upon investigation I found that there are forty-three species in the genus *Autographa* of which at least fifteen are recorded in North America, and some of these – particularly *A. californica, A. bimaculata, A. buraetica, A.*

*pseudogamma, A. v-alba* and *A. corusca* – are not only found in North America, but bear a remarkable resemblance to the Eurasian Silver Y.

The editorial of the latest edition of the *Bulletin of the Amateur Entomological Society* notes that there has been somewhat of a population explosion in this species during 2010, and in insects, as with people, a population explosion often triggers a diaspora.

I would originally have placed a fair amount of money upon the moths pictured here being the Silver Y species with which I have been familiar for four decades. However, now I am not so sure. I would be very interested in reading the comments of any expert in the North American Noctuidae.

### REFERENCE

*Cirsium Palustre* (Marsh Thistle) literature search and habitat potential risk analysis (Nancy Fraser, Canadian Ministry of Forests, Vancouver, 2000)

# THE MYSTERY ANIMALS OF IRELAND

## GARY CUNNINGHAM & RONAN COGHLAN

**Available now £8.99/$US15**

# EXCLUSIVE EXTRACT

One of the most entertaining things about running a publishing company, especially a publishing company that is free from the constraints of having to make a profit, is that I get to commission books that I would personally want to read.

As regular readers of my inky-fingered scribblings both here and elsewhere may well know, we have been publishing a series of books on the mystery animals of the British Isles, which contain exactly what it implies on the tin.

Although, geographically speaking, Ireland is part of the British Isles, politically – since 1922 – it has not been, and, having no wish to offend anyone living in the Emerald Isle, I decided to commission a separate series on Ireland's many and peculiar mystery animals.

Both Gary and Ronan are fine researchers and have been friends of mine for some years, and so it was with great pleasure that I commissioned them to write this book, knowing full well that it would turn out to be excellent, which it most certainly did!

Enjoy! JD

# 15

# The Dwarf Wolves of Achill

I first learned of the Dwarf Wolves of Achill Island whilst reading the September 1997 issue of *Uri Gellar's Encounters.* (issue no. 12). An article by Dr Karl Shuker entitled *From Liver Birds to Leopard Cats* delves into the mythological animals of Britain and Ireland. On page 22 is a reference to the Achill wolves.

*According to Irish tradition, Co Mayo's Achill Island was home to a*

> *type of small wolf-like beast long after true wolves died out elsewhere in the British Isles. They were said to resemble normal wolves in overall appearance, except for their small stature.*

Even though I had been aware that my own country (in the more remote regions at least) was said to contain unidentified lake monsters, I was especially thrilled that there was yet another creature that remained unaccepted and undiscovered by the zoological community.

When Dr Shuker's classic book *Extraordinary Animals Revisited* was republished by the Centre for Fortean Zoology in 2007 on page 95 I found it mentioned once again the Dwarf Wolves of Achill. However, on this occasion, he provides the reader with a tantalising new facet of these enigmatic creatures

> *In a letter to me of 21$^{st}$ February, 1998, British zoologist Clinton Keeling provided me with a fascinating snippet of information on this subject – revealing that as comparatively recently as 1904 the alleged Achill Island wolves were stated to be "common" than no less a person as Sir Harry Johnston, discoverer of the okapi.*

One of my regrets is not contacting Clinton Keeling in order to elicit more details. I can't help but wonder about the origins of the exact year he quotes and, of course, how Sir Harry Johnston discovered the mysterious canids in the first place.

## Achill – the island of bogs and cliffs

Achill is Ireland's largest offshore island (280 square metres). It is certainly an island of superlatives. It possesses the lowest and highest corrie lakes in the country. Even more spectacular are the vertiginous cliffs of Croaghaun, which form a 3km precipice on the island's north-eastern end. They drop an incredible 664m/2192 feet sheer into the Atlantic below and as such are more than three times the height of the famous Cliffs of Moher in Co Clare. In fact, they are the highest sea-cliffs in Europe. Achill also has the Menawn Cliffs which extend for approximately 2km and rise to more than 450m/1350 feet for most of their length.

Even though Achill has many awe-inspiring geographical features, its wildlife is sparse to say the least. However, this wasn't always the case, will early naturalists or commentators mentioning the proliferation of animal and bird life in particular. In fact, both the golden and white-tailed sea-eagle existed in astonishing numbers with successful breeding occurring up to the end of the 19$^{th}$ Century. Its varied bird life included chough, snow bunting, ring ouzel, grouse and other raptors such as peregrine falcon, kestrel, merlin and sparrowhawk. Sea birds included razorbills, guillemots, puffins, cormorants, northern divers and gulls. Tragically, the bird life has been persecuted to such an extent that the only places where it can be seen today in ant numbers are the inaccessible cliffs of the island's northern side.

Significantly, certain terrestrial mammals such as foxes, hares and rabbits were also very common at one time. It is important to remember whenever I visited Achill in April, 2009, the only wildlife we saw were some seagulls, sheep, swans and a hare bounding across the road at

Dooagh village. Compare this with the deafening sounds of the 50,000 plus seabirds, golden hares (a unique subspecies), buzzards and 30 or so grey seals we watched one day on Rathlin (Co Antrim)

## Other mini-wolves

If there are Dwarf Wolves on Achill, it might be interesting to see if other populations of such diminutive creatures have evolved. Intriguingly, Mother Nature has given us several other forms of dwarf wolf on a couple of occasions. The most famous is the Japanese shamanu (*Canis lupus hodophilax*) which was (and possibly still is) the world's smallest species of wolf, incredibly only 14m at the shoulder and approximately only 45m in length. The reason for its diminutive height is due to its unusually short legs – in fact, this morphological feature was so destructive that some zoologists assigned the shamanu full species status. Even though this miniature wolf was revered by the Ainu people, a group of indigenes who preceded the main Japanese population to the islands, as "the howling god" it was still persecuted by hunting and deforestation. Tragically, the shamanu was declared officially extinct in 1905. However, sightings of small wolves continued after this date with the most striking evidence coming in the form of nineteen photos taken by Hiroshi Yaqi in 1966.

Another miniature wolf was the Falklands Islands wolf (*Dusicyon australis*), which was also vertically challenged, reaching a height of 24m at the shoulder. As with all wolves, the world over, it too was hunted to extermination by early Spanish and Scottish settlers. Sadly, by 1876 it was no more and yet another form of wolf was extinct.

The Hungarian reedwolf was a mysterious and as yet uncategorised canid reported from Hungary and eastern Austria up to 1900. Its identity has never been satisfactorily determined and yet there are specimens in museums around the world. It may even be, not a wolf, but a jackal.

Before leaving the subject of mini-wolves, it is important to note that there were some prehistoric forms. One such example was the small race of wolf indigenous to the Channel Islands off the coast of California. This wolf certainly evolved to take advantage of the abundance of prey, in particular the dwarf mammoths (*Mammuthus exilis*) which were themselves only 6'/1.8m in height.

From the previous accounts it can be seen that the wolf, even though it has suffered considerable assault by man, has adapted and evolved to survive in many habitats. Even more thought-provoking is its occurrence on islands, showing that the alleged Achill wolves would not be unique.

## Why are islands zoologically special?

Islands have always exhibited a certain mystique with regard to their ecology and wildlife. In fact, zoologists recognise a process whereby evolution affects the size of island mammals. This rule is known as Foster's Rule and demonstrates that smaller animals tend to increase in size, while larger ones tend to decrease. The reasons for this are relatively straightforward – the smaller mammmals such as mice have less predators to worry about so they can evolve into larger island varieties, e.g., the St Kildan field mouse (*Mus hirtensis*). Conversely the larger mammal species are restricted by the availability of food and consequently over generations reduce their body size to cope with this fundamental problem.

Significantly, from an evolutionary point of view, why would the European wolf evovlve a smaller version on Achill Island in the first place? As previously mentioned, the island did produce sufficient prey, but the lack of forestation for cover would definitely have been a major problem.

## What exactly were Achill's dwarf wolves?

It is worth mentioning here that in Shuker's *Extraordinary Animals Revisited* there is a very intriguing clue to the identity of Achill's dwarf wolves. The possibility that young wolves or even jackals were imported for hunting is very plausible. These small wolves allegedly existed until 1920. The last normal wolf in Ireland was said to have been killed in 1786, though this date has been disputed.

In 2007 I became aware of another possible identity after discussing the subject with my friend and correspondent Pap Murphy.. When I asked him about Achill's dwarf wolves he immediately spoke about the *cordog*. This was a medium-sized dog which apparently was half collie and half terrier.. Pap also informed me that it was fast, ferocious, clever and was used to hunt rabbits at night time. Significant was the once common belief on Achill that wolves and dogs bred (which they can do freely) and the result was the previously mentioned cordog. I am currently trying to obtain a photo of a cordog as the breed is sadly extinct. Therefore, this is possibly the only way to ascertain if there was any morphological relationship with the wolf and, if so, would be the only evidence thus far to help prove the one-time existence of the Achill miniwolves.

The only other possibility to account for the dwarf wolves would be confusion with foxes. (Intriguingly, at one time on the island foxes were apparently tamed to hunt rabbits). I find this difficult to understand how the ubiquitous fox would be confused with the much-maligned wolf, no matter how large or unusual the fox was.

## Now for the ambitious part! Puzzling aspects.

Despite the lack of concrete evidence for this creature's very existence and, indeed, the lack of reports after 1905, I do feel there is a genuine unidentified animal responsible for the dwarf wolf tag. As cryptozoologists are well aware, one of the main obstacles to mainstream scientists' accepting their work, including their research and conclusions, is their over-reliance on anecdotal evidence, i.e., eyewitness accounts. Although there are some good sightings regarding dwarf wolves there are also a few niggling fundamental questions which need to be answered. For example, if there were genuine/bona fide dwarf wolves on Achill, how did they come to be on the island in the first place? Why have no archaeological remains been found to date, considering the amount of digs which occur on the island?

Whenever I started to research the factors which would be required for a population of wolves, dwarf or otherwise, to survive on Achill, I was immediately struck by the lack of habitat in the form of forestation. Crucially, how would a small population not succumb to the problems of disease due to a restricted gene pool as a result of so small a population. How many dwarf wolves existed in Ireland in the first place?

## Final thoughts

Frustratingly – perhaps tellingly – Achill's dwarf wolves weren't mentioned by any of the prominent naturalists of the 19th Century, such as Scharff, Stelfox, Barrett or Hamilton. Robert Llloyd Praegar doesn't mention the possibility in his seminal work *The Way That I Went* and, if they were supposedly common until 1900s, this would have been expected. Previously I mentioned the lack of suitable forested areas on Achill. Such would be of course a necessity not only for more effective hunting, but to conceal their dens safely away from other predators and, of course, man. However, there is a forested area on the mainland north-east of Mullarany known as Cartron. Intriguingly, if this area was forested until 1900s, then conceivably a population of dwarf wolves could have traversed Achill Sound at low tide, looking for opportunities on Achill. There is also the possibility that the animals from the mainland were normal-sized Irish wolves and over generations became smaller to adapt to the prey and lack of cover present on the island.

Out of all Ireland's undiscovered animals, the possibility that a form of miniature wolf existed on the country's largest island (which is admittedly small by Scottish standards) some 130 years after the last remaining wolves became extinct is definitely very exciting. How eerie it must have been for Achill's inhabitants to have heard a howl in the dead of night over the incessant sound of the Atlantic crashing at the headlands and base of the precipitous cliffs on Europe's most westerly land.

As I have been unable to discover any newspaper references to dwarf wolves on Achill Island it might be assumed that the belief that they survived to the incredibly recent date of 1910 must have been due to a mistake or another animal altogether.

Dr Kieran Hickey of University College, Galway (National University of Ireland) has conducted extremely thorough research on all aspects of the history of the wolf in Ireland. In fact, it is probably the most extensive study of this much maligned apex predator in the Emerald

Isle. Dr Hickey has a book forthcoming in 2010 which I am eagerly awaiting.

He has informed me that it is very unlikely, almost impossible, that any wolves, whether stunted or regular-sized, survived beyond 1800. The presence of suitable habitat large enough to sustain a genetically viable breeding population free from inbred illnesses coupled with adequate food is lacking. However, he is open-minded concerning the possibility of other as yet undiscovered animals in the remote, underpopulated wilderness areas in Ireland. He believes such animals are individuals that are very distinctive morphologically in one way or another, e.g., in colour, shape and particularly size. For example, he believes the legend of the Dobharchú is based on eyewitness accounts of extremely large individuals within regional populations of the Irish otter. After many years of investigation of this creature, I would be inclined to agree with that proposed identification.

In conclusion, it seems as though regrettably we have reached a dead end in respect of the survival of a dwarf form of wolf on Ireland's largest island – Achill. Unless any archaeological remains are discovered in the future which would conclusively prove its existence, then Achill's vertically challenged wolves will have to remain, as is so often the case in cryptozoology, an enigma – albeit a very fascinating and intriguing one.

# OLIVE MILLIPEDES

## Lucy Henson

Olive millipedes (*Spirostreptus gregorius*) are a commonly kept millipede in the hobby as they are very active and, compared to other species of millipede, they are easy to rear. This particular species is docile and generally unwilling to release its defensive chemicals whilst being handled. The chemicals vary both in intensity and its effects between species, and it is often advised that millipedes should not be handled unless absolutely necessary. Most of the common species that occur in the hobby are, however, relatively harmless, but there are exceptions. These secretions may cause an allergic reaction and repeat exposure has been linked to cancer; therefore, if you do deem it necessary to handle your millipede, always wash your hands afterwards and avoid getting the secretion in your eyes and mouth, where it can do the most damage.

It seems pretty obvious where they get the name 'olive millipede' from once you obtain one of these fantastic creatures. Their bodies are a lovely olive colour and have black lines that indicate the separating segments. Their legs have an orange shade, as do their antennae. Males and females appear identical, and both grow roughly to around 8-11cm, which means that like many millipedes there is no easy way to sex this species without searching, so to speak. As a rough indication, males appear to be missing a set of legs on the seventh segment, whereas females do not. This can be checked by placing the millipede in a clear box and looking at the underside. Adult females should also appear to be slightly chunkier than males, but this is only notice-able with a large group of adults.

As for housing these guys, it is pretty much the same as most other species; the tank is advised to be three times the length of the millipede's body, but the larger the better. The best substrate is a mixture of rotting wood and dead leaves, just as you would use for fruit and rhino beetle larvae. The wood is advised to be rotten to the point of which it easily crumbles in your hands and is generally white in colour as the fungus bleaches the wood. Soak the substrate in a container full of water overnight, in order to kill off parasites and other insects, and soften the substrate. Substrate should be at least ten cm thick and should always be damp, but not wet. It may have to be left to dry out after soaking for a couple of days in a lidless tank. When it comes to feeding, millipedes are primarily vegetarians. Dandelions, carrot, parsnip, cucumber, mango, apple etc are good examples of foods millipedes will love. They live on mainly decomposing plants and as their exoskeletons consist partly of calcium, they need a supplement of some sort.

More often than not, a varied diet offers this calcium but leaving a cuttlefish bone in the tank is never a bad thing.

When I used to keep millipedes I found the olive millipede a delight to watch and although it isn't usually recommended to house millipedes together, these did really well. If you are thinking of housing a group together, I would recommend sticking to one species and making sure the enclosure offers sufficient space. As for pricing, olive millipedes are not that expensive: you can buy large juveniles for around £2.50 each and adults for around £10 each. They will breed well providing you start with a group of both sexes and large colonies are often produced, which require fairly large tanks. Due to this, there is no place in the hobby for wild-caught olives, and you should always check yours are captive-bred. These may not be the most sought-after species of millipede;

however, they are an easy species for beginners and will allow for mistakes so you can first gain experience if you wish to aim for larger and trickier species in the future. The fact that they are an easy beginner should not put you off; these are great fun to watch and as they are an active species they make a great display species.

"Though here olive millipedes are listed as *S. gregorius*, there is a lot of confusion regarding what their real scientific name is. Over six wildly different names have been used for them and no-one really knows which one is correct, though the name used here seems to have the majority usage. It may be best to refer to them as Spirostreptida sp. "olive", which, because Spirostreptida is a huge order of millipedes, is equivalent to calling a lion "Carnivora sp. "furry""..."

© 2007 Sarah Fowler

# MYSTERIES OF THE DOG

## Scottie Westfall

A zoological mystery lurks in many Western households. Although we live intimately with the species associated with this mystery, only the contours of the possible answer to this conundrum have been revealed.

And each time new studies reveal new information, we become more and more confounded. New information merely confuses us, distorting the paradigms through which we have assumed to be true.

The zoological mystery I am talking about here is something many of us have not fully considered: What exactly is a domestic dog?

Some parts of this question can be answered. It is a member of the order Carnivora, which includes a whole host of mammals that range in size from the tiny least weasel to the enormous southern elephant seal. We also know that dogs are part of a larger family called the Canidae, which includes the foxes, the wolves, the jackals, the coyotes, and the many unusual South American wild dogs, such as the maned wolf and the short-eared dog. We also know that domestic dogs belong to the genus *Canis,* which includes the wolf, three species of jackal, the coyote, and the Ethiopian wolf or Simien jackal. Depending upon how one views the exact taxonomy of the wolf, this genus also includes the red wolf, the Eastern timber wolf of North America, the Himalayan wolf, and the wolf of the Indian subcontinent. The exact taxonomic status of these four wolves is unclear and hotly contested, for they may represent distinct species, rather than belonging to subspecies of the true wolf, *Canis lupus.* (I prefer to use the term "true wolf" rather than grey wolf, simply because the actual species comes in more colours besides grey.)

Where the domestic dog fits into the genus *Canis* is the big problem. Darwin believed that domestic dogs were the result of mixing several different wild canids. Dogs can interbreed and produce fertile offspring with golden jackals, coyotes, and all animals that are either part of *Canis lupus* or very closely related to it. Konrad Lorenz, the Austrian ethologist, also thought that dogs were derived from more than one wild ancestor. Lorenz contended that the sociable and very trainable breeds of domestic dog were derived from the golden jackal (*Canis aureus*).

The more reserved breeds that were difficult to train and bonded very strongly with one or two people were derived from wolves. He divided domestic dogs into two distinct types, which he referred to as "aureus dogs" and "lupus dogs."

However, various behavioural, genetic, and archaeological studies have revealed that the ancestor of the domestic dog is most likely the wolf. In 1993, geneticist Robert Wayne (who is now at the University of California—Los Angeles) published a study in the journal *Trends in Genetics* in which he compared mitochondrial DNA (MtDNA) samples from a variety of canid species. He found that domestic dog MtDNA was most similar to *Canis lupus.* Later studies of nuclear DNA from dogs and their relatives—including one genome-wide study—found that dogs and wolves were much more closely related to each other than to any other species.

And these differences were miniscule.

The scientific consensus is that dogs are derived from the grey or true wolf and should be placed within the species *Canis lupus.* When considered part of this subspecies, the dog is called *Canis lupus familiaris.* And because the wolf is the wild ancestor of the dog, most research on the ancestry and origin of the domestic dog focus upon that species.

Of course, a minority opinion still exists. Janice Koler-Matnick, a researcher who has expertise in the New Guinea singing dog breed, contends in a 2002 article in the journal *Anthrozoö* that the dog is actually derived from another species besides *Canis lupus*. It was very closely related to the wolf, but it was much more generalized in its diet and quite a bit smaller. She calls this ancestor the wild *Canis familiaris,* using the old Linnaean name for the domestic dog. She pays very close attention to the fact that dogs and wolves do not share a wide range of DNA sequences. If the two were the same species, they would share many different MtDNA sequences.

However, this theory has generally been rejected in the light of even more recent genetic analysis. Koler-Matznick's view that dogs were derived from a more generalist ancestor

than the wolf of Northern Eurasia does not require a separate species at all, and this more recent genetic evidence points to what I shall call the "forgotten wolves."

It is well-established that there are two basic clades within the species *Canis lupus*. The best-known of these clades is the northern clade. These are the hyper-carnivorous wolves of Northern Eurasia and northern and western North America. These are big wolves. Some exceptional individuals from Alaska have exceeded 140 pounds in weight, and they all possess massive heads in proportion to their body size, powerful jaws, and large teeth. They need these adaptations to kill large ungulates, like moose, bison, and, at one time, the great aurochs that once roamed Europe.

These are the wolves most familiar to Western-

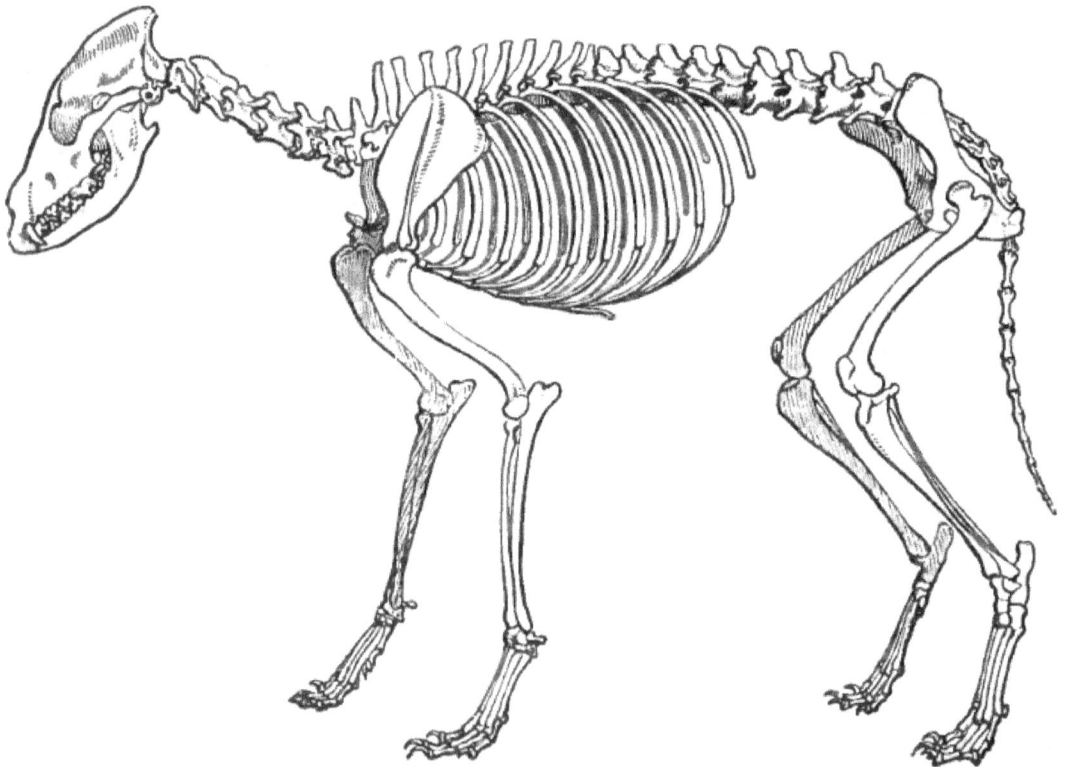

SKELETON OF WOLF.

ers. They are the ones that appear in movies. They are the ones that are featured in zoos and wildlife parks. These are the animals that come to mind when the word "wolf" is mentioned.

All recent wolves living in Europe have been of this clade. The main wolf subspecies of Europe is *Canis lupus lupus* or the common wolf. Its range once ran from Ireland and Britain to France and then all the way east to China, including the southern part of China.

However, the original wolves that lived in Europe as recently as 500,000 years ago were not like these animals at all. They were more slightly-built animals that possessed smaller heads and less powerful jaws. This animal had to have been a generalist in much the same way Koler- Matznick describes the dog's ancestor. After 500,000 years ago, the wolf of Europe, Northern Asia, and China evolved into the big game hunting wolves that needed the specialised adaptations we see in the modern wolves of the northern clade.

Not all wolves experienced this change.

The wolves of India and the Middle East retained the more gracile features we associate with the ancestral wolf. The southern clade wolves have smaller brains and heads and possess less powerful jaws. It is often suggested that the Mexican wolf (*C. l. baileyi*) has some of these southern clade features. If the Eastern timber wolf (*C. l. lycaon*) and the red wolf (*C. rufus* or *C. l. rufus*) are actually derivatives of the Old World wolf species, then they, too, have retained some of these primitive features. Of course, some of these traits can come out through hybridisation with the coyote.

Because analysis of Native American dog MtDNA points to an old world origin for the domestic dog, we know that we have to look at Eurasian wolves for their possible ancestors.

Dogs have smaller heads and brains than wolves do. They also generally have less powerful jaws. It makes sense that we would look

to the southern clade for the ancestral wolves that gave us dogs.

Within the Old World southern clade, a taxonomic issue exists. The wolf of Anatolia, the northern part of the Levant, Mesopotamia, Iran, and Afghanistan is often considered the same subspecies as the wolf from the Indian subcontinent. This subspecies is called *C. l. pallipes,* the pale-footed wolf. However, the wolves of the Indian subcontinent have a much older MtDNA sequence than the other wolves counted in this subspecies. Its exact status is unclear, but it may represent an entirely different species (*Canis indica*), which would mean that it cannot be considered a dog ancestor.

However, the other wolves that have been placed in *pallipes* are still in consideration.

Why?

Well, in March of this year, a team of researchers at UCLA released the findings of genome-wide study of dog and wolf DNA. The study looked at 48,000 different parts of the genome to compare variance between domestic dogs and extant wolf populations. The study found that the primary source for genetic material in the domestic dog were wolves from the Middle East. The researchers used samples from wolves from Iran, and the only wolves that live in Iran are *pallipes.*

The other subspecies likely included in their sampling of Middle Eastern was *C. l. arabs*, the Arabian wolf. The sampling included wolves from Saudi Arabia and Israel. The only wolves in Saudi Arabia are Arabian wolves, while the northern Israel has a *pallipes* population and southern Israel and the Sinai have Arabian wolves.

These two subspecies are likely the most closely related to the ancestors of the domestic dog. The Arabian wolf is the smallest living subspecies. It is roughly the size of a dingo, and in its conformation very strongly resembles this Australian dog more than any other wolf. The *pallipes* wolf is a bit larger, roughly the size of a golden retriever or a Labrador. Its coat is often ginger or tawny in colour, which is trait that is quite common in domestic dogs.

Of course, this study is far from the final word. A competing hypothesis on the dog's origins comes from a comparison of the diversity in MtDNA haplotypes in domestic dogs throughout the world. Greater variance in MtDNA sequences is associated with older populations.

The population with the greatest variance is generally going to be the oldest population. Peter Savolainen, a researcher at the Royal Institute in Stockholm, Sweden, found that the region with the most diverse MtDNA was in East Asia. Using Savolainen's methodology, a Chinese team pinpointed South China as the location for the first domestic dogs. If that finding is true, then the dog's primary ancestor would have been the common wolf, *Canis lupus lupus*. And dogs would have derived from the northern clade of wolves.

Such a thing is not impossible. One of the places with the oldest dog remains is in Israel. The so-called Natufian culture was among the first people start small-scale agriculture and live in villages. The remains of dogs have been found in their sites that have dated to 12,000 years ago. Tamar Dayan, a zoologist at Tel Aviv University, compared the remains of these wolfish dogs to the remains of *pallipes*-type wolves that were living in the area during the same time period. The wolves were significantly larger. Indeed, they were even larger than modern *pallipes* wolves that live in Israel.

So, it could be possible that dogs could have been domesticated from one of the northern clade wolves.

However, it would be nice if we could be sure that these 12,000-year-old dogs were among the first domesticated, but we simply do not have that luxury.

When dogs were first domesticated is as con-

tentious as where they were first domesticated. The oldest accepted dog remains come from Russia and date to about 15,000 years ago. Another dog was found at Obercassel in Germany. It has been estimated to be 14,000 years old. However, a very controversial skull was found at Goyet Cave in Belgium. It was recently dated to 31,700 years old, which would place dog domestication within the Aurignacian period of Europe.

As was noted earlier, the more we find out about our dogs' past, the more murky things become. It is confusing that dogs would have the greatest variance in MtDNA sequences from Southern China, yet have greatest genetic similarities with Middle Eastern wolves. And it gets even more confusing when the oldest possible dogs skulls all come from Europe.

How is this all possible?

Well, if one carefully reads the study that shows that suggests that dogs are derived from Middle Eastern wolves, one sees something unusual. The dogs from East Asia have a stronger than normal genetic similarity with Chinese wolves. The researchers suggest that it could mean that dogs in East Asia were heavily interbred with wild Chinese wolves.

Or it could mean that there was a separate domestication event. The Chinese wolf genes entered that population through mixing dogs derived from Middle Eastern wolves with an indigenous East Asian dog population. It could also mean that the Goyet dog was the result of another domestication in which the entire genetic base of the original population was replaced.

The MtDNA of the Goyet Cave dog and the other wolves and dogs from Pleistocene Europe has been compared to MtDNA of modern dogs and wolves. There is no match. That means that this dog did not contribute to any living dogs.

However, it still does not mean that it is not a dog. It could mean that modern dogs are a replacement of that aboriginal European dog. Perhaps there was something about dogs derived from *pallipes* and *arabs* that made them more fit to live with people. Perhaps they were just hardier.

This hypothesis is worth considering.

If this scenario is true, then we do not need to throw out the Goyet Cave dog, just because it appears to be "too old." It is likely that Paleolithic peoples did form an alliance with wolves, and they may have done so at many different locations in Eurasia. Contrary to what some may assume, hunter-gatherer people are often known to make pets out of wild animals.

Wolves that have not experienced persecution and live in prey-rich areas are actually quite curious about people. (Accounts of them doing so in the Old West can be found in Bruce Hampton's *The Great American Wolf* and Mark Derr's A *Dog's History of America*, both of which will be listed in the bibliography.) Attempts to domesticate canids has occurred in other places besides Eurasia. The people of Tierra del Fuego domesticated the culpeo, a type of South American "fox," which, like all South American foxes, is actually more closely related to dogs and wolves.

So it would have made perfect sense for Paleolithic to try to domesticate these animals.

Wolves would have helped in the pursuit of large prey. They would have been excellent in the pursuit of deer, and they may have even helped people hunt mammoths. Carl Lumholtz, the Norwegian ethnographer, described indigenous Australians taking dingo pups from dens. These pups grew into adults, which were then used to help them hunt prey. It is possible that the first wolves were domesticated in this fashion. So helping in the hunt was likely an early function for these first dogs.

We do not know.

But as we look deeper and deeper into the past

of our dogs, we realise exactly how much of a mystery they really are. They are four-legged enigmas that seem so common, so well-known. How could they possess so many mysteries?

But the question still remains: What exactly is a domestic dog?

# Bibliography

Dayan, T. 1994. Early domesticated dogs of the Near East. *Journal of Archaeological Science,* 21, 633-640

Derr, Mark. *A Dog's History of America: How Our Best Friend Explored, Conquered, and Settled a Continent.* 2004. North Point Press.

Germonpre, M. *et al* 2009. Fossil dogs and wolves from Palaeolithic sites in Belgium, the Ukraine and Russia: osteometry, ancient DNA and stable isotopes. *Journal of Archaeological Science.* 36, 473- 490.

Koler-Matznick, J. 2002. The origin of the dog revisited. *Anthrozoös* 15, 98-118.

Lindblad-Toh, K. *et al* 2005. Genome sequence, comparative analysis, and haplotype structure of the domestic dog. *Nature.* 438, 803-819.

Lorenz, Konrad. *Man Meets Dog.* 2002: New York, Routledge. Originally published in German in 1949 under the title: *So kam der Mensch auf den Hund,*

Pang, J. *et al* 2009. MtDNA data indicate a single origin for dogs south of Yangtze river,lessthan 16,300 years ago, from numerous wolves. *Molecular Biology and Evolution.* 26, 2849-2864.

Savolainen, P. 2002. Genetic evidence for an East Asian origin of domestic dogs. *Science.* 22, 1610-1613.

vonHoldt, B. *et al* 2010. Genome-wide SNP and haplotype analyses reveal a rich history underlying dog domestication. *Nature.* 464, 898-902.

Wayne, R. 1993. Molecular evolution of the dog family. *Trends in Genetics.* 9, 218-224.

**CFZ PRESS**

www.cfz.org.uk

I t has always gone somewhat against the grain to print reviews of our own books within the pages of our publications. Even when we gave them to an impartial third party to review we always felt that it wasn't quite cricket, and that it was the sort of thing that a chap knew full well that other chaps might feel that it was something that a good chap probably wouldn't do. So, starting from now, we will be including a brief rundown of recent books published by CFZ Press in each edition of our two journals.

We have another problem. From 2004, when we started publishing books through Lightning Source, until very recently we had always included a voluminous listing of all our publications, complete with ISBN numbers, a brief description of the contents, and a thumbnail picture of the front cover. When – earlier this year – our catalogue passed 60 titles, this began to be increasingly cumbersome, and so we did away with that.

Another recent change to CFZ Press was the launching of our second imprint – Fortean Words. This is for stuff which is neither natural history, or cryptozoologically based, but still fits within the Fortean framework of sub-

jects of interest both to us, and to the core demographic of people to whom we hope that we continue to appeal. We launched this new imprint with several UFO-related titles including the start of a multi-part encyclopaedia of British UFOs, and veteran Fortean Andy Roberts' own inimitable take on the Berwyn Mountains incident in North Wales during 1974.

Since the last issue we have published volume four of our collected editions of our cryptozoological journal *Animals & Men.* This covers the years 1997 to 1999. We have also published Richard Freeman's magnum opus on Japanese monsters in folklore, myth and media. We believe that *The Great Yokai Encyclopaedia* is the most important book of its kind on the subject and we have high hopes for it.

We are also very proud to announce the publication of two books that have been in the works for a long time. They both are the result of inspired collaborations between old friends of the CFZ family, and are both excellent reads.

*The Mystery Animals of Ireland* by Gary Cunningham and Ronan Coghlan is previewed elsewhere in this edition of *The Amateur Naturalist.*

The other volume, *Monsters of Texas,* is the result of a collaboration between British Fortean researcher Nick Redfern who now lives outside Dallas, and native Texan Ken Gerhard, and covers a wide range of the mystery animals, both corporeal, and less so of the Lone Star State.

Our final new book this time around is *Tetrapod Zoology Book One* by Dr Darren Naish. For some years he has been writing a highly acclaimed blog called Tetrapod Zoology in which he explores some of the more arcane aspects of – you've guessed it – the zoology of the tetrapods. For the first time articles from the early version of this blog now appear in book form and we are very proud that it is us who have published it.

## LUCY'S LIFE

Lucy is a young lady of eighteen, who appreciates the important things in life; she breeds mantids, has a bevy of escaped hamsters under her bath, and has a little brother who is obsessed with snails. In short, the perfect columnist for *The Amateur Naturalist*

Hello again! It might be just me, but is this year going extremely fast?! It is nearly the September Hamm show in Germany, which I was hoping to get to however have been unable to get a passport sorted which is a real bummer, therefore it's the usual "Maybe next year". Hope it's a good one, if any of you are going! I love looking at the pictures, seems like a great event!

First off I bring sad news, Leonardo, my mack snow leopard gecko, whom you may have read about last issue, has unfortunately passed away. He was nearly 2 years old; rest in peace little Leo. I am not sure whether I will be getting another Leopard gecko, perhaps in the future, however now I'm fully devoted to my snakes.

At present I only have 2, William the cornsnake and Jim the Royal Python, however I plan to change this very soon as I'm looking for a hognose to add to the collection. Now, I've heard a few horror stories about these guys biting and not only refusing to let go but also chewing… This does scare me a little however I'm hoping I'll be able to find a good breeder and get a nice tame one!

Even though I love snakes, my passion for inverts has not totally gone. I recently received some assassin bug, *Platymeris biguttatus,* eggs from a good friend and they have all hatched! I haven't been sprayed or bitten and none of them have escaped – touch wood - but I find these little guys fascinating! They have a black head, red body and yellow legs, which looks great on display!

They are primarily fed on mealworms of various sizes however I am yet to witness them feed, one day I will catch them in the act! Ignore all the horrible stories about these guys being nasty, they are great. As long as you're careful you have nothing to worry about. I mean if I can keep them and not get hurt!

Even Alex (the snail lover) likes these guys; I will admit he did believe me when I said they have little guns like real assassins do… But the colours are amazing! (Eds. note: Assassin bugs were covered in issue 3 of *Exotic Pets*. He would also like to remind everyone to look out for *Platymeris* sp. "Mombo" because they are truly amazing…)

Another show I was hoping to attend this year was the second IHS Doncaster show. I hopefully will be attending, but am not 100% sure. The last time I went, in June I think it was, I managed not to come home with anything?! -Which was a big shock to my family and to myself if I'm honest… Hopefully I'll make it up for the next show! I'm thinking of maybe getting a few inverts and might accidently come home with another snake… So watch this space!

To wrap it up, I hope this year is living up to all the expectations everyone had for 2010! We are already $3/4$ way through and I don't mean to scare you but Christmas will soon be upon us! And on that bombshell, thanks for reading!

## corinna's endangered species column

The 2008 IUCN (International Union for the Conservation of Nature) Red List includes the Philippine Eagle (*Pithecophaga jefferyi*) states that there are only around 300 of this critically endangered bird of prey surviving in the wild in the Philip-pines. Also known as the great Philippine eagle or the monkey-eating eagle, it is one of the tallest, largest, most powerful and rarest birds in the world and belongs to the family Accipitridae. In certain areas of the Philippines it is known as Haribon, or Haring Ibon which means 'Bird King'. It became the country's national bird in 1995 by virtue of Presidential Proclamation.

The Philippine eagle is approximately 1m in height, has a wingspan of 2m and weighs around 6.5kg. It has been shown to be an opportunistic hunter but its main diet includes flying lemurs, snakes, civets, hornbills, some squirrels, and sometimes bats and monkeys. This eagle is known to live for about 30 to 60 years.

One of the main reasons for their threatened existence in the wild is deforestation by logging and expanding agriculture – old forest is being

lost at a high rate and in the lowlands the eagles' habitat is owned by logging companies. Other threats come in the form of poaching, mining, pollution and even exposure to pesticides that affect breeding. Sometimes they are found in traps that have been laid by local people to catch deer. Killing this critically endangered species is punishable under Philippine law by twelve years in jail and heavy fines. There was also a reduction in numbers caused by the eagle being caught for zoos, although this is no longer an attributable cause.

Unless there is intervention, this magnificent bird will disappear completely from the wild. In recent years areas of land have been declared as protected specifically for this bird – the 170,000 acre Cabuaya Forest and the 9,200 acre Taft Forest Wildlife Sanctuary on Samar for example. There is also ongoing research into the eagle's behaviour, ecology and population dynamics being carried out by The Philippine Eagle Foundation in Davao City, Mindanao, which is dedicated to the protection of the species and its habitat. This foundation has successfully bred Philippine eagles in captivity for over ten years and has conducted the first experimental release of a captive-bred bird into the wild. The foundation has 32 eagles at its centre, of which 18 were bred in captivity.

The world famous Charles Lindbergh – who solo-crossed the Atlantic non-stop in 1927 became fascinated by this eagle and in his capacity as representative for the World Wildlife Fund, he travelled to the Philippines several times between 1969 and 1972 and helped to persuade the government to protect the bird. In 1969 the Monkey-eating Eagle Conservation Programme was born to help preserve this species and in 1992 the first birds born in captivity were through artificial insemination. It was not until seven years later that the first naturally bred eaglet hatched. The first captive-bred to be released in 2004 unfortunately died by accidental electrocution in 2005 and another released in 2008 was shot and eaten by a farmer. http://philippineeagle.org/

# BOOKSHELF

*The Naturalized Animals of Britain and Ireland*
By Christopher Lever
**Hardcover:** 424 pages
**Publisher:** New Holland Publishers Ltd (30 Oct 2009)
**ISBN-10:** 1847734545
**ISBN-13:** 978-1847734549
**Product Dimensions:** 23.8 x 15.6 x 3.8 cm

Once upon a time, when the world was young (or at least when I was), I went into Bideford on the bus one Saturday afternoon. At the time I was a spiky haired young Herbert, agog to hear what Johnny Rotten's new band was going to sound like, but then, as now, I was an obsessive animal naturalist. I went into the excellent book shop which for so many years was one of the features of Mill Street - and which I am sad to say like so many things from my childhood and adolescence - no longer exists, and bought a remarkable book. It was called *The Naturalized Animals of the British Isles* by Sir Christopher Lever who has been one of the world's leading experts on the subject of (you guessed it) naturalised animals for nearly half a century. He is now 78 and his various books have become sacred texts to me, Max and other core-members of the CFZ for the past 30 years. I have waited since that day in 1977 to get my sticky little fingers on his follow-up.

This afternoon there was a knock on the door and, someone having forgotten to shut one of the connecting doors to the rest of the house, Biggles my dog rushed up and nearly knocked the surprised delivery man over, head-butting him in a part of the anatomy which he (Biggles) no longer possesses, but which caused me (and will no doubt cause anyone reading this who is in possession of a Y chromosome) to wince. Doing my best to placate the poor chap I got rid of him as soon as I could, and ripped the packaging open. This was a book I had waited 33 years to read.

I suppose, like any long-awaited cultural artifact or milestone, it could not really fail to be a disappointment. Max and I wanted far too much for it and I am afraid we were disappointed.

I feel very churlish criticising this latest magnum opus from such an important and pivotal figure in my intellectual life. However, it is actually his fault. With the first edition of this book, and with subsequent books like his remarkable *Naturalized Fishes of the World* (Academic Press Inc (Sep 1996)) he raised the bar so impossibly high that no one volume aimed at a general audience could ever provide what specialists like the editorial team of this publication would have wanted. For in the intervening three and a bit decades since the spiky haired 18-year-old Jonathan Downes first avidly devoured the original edition of this book, I have not only lost the spiky hair, and the callowness of youth, but I have spent most of the intervening years studying a wide range of zoological subjects and particularly the naturalised animals of the British Isles. I have also written widely on the subject, and I am disappointed to find that various things that I discovered, and helped to discover, such as what appears to be a population of the wels catfish

*(Silurus glanis)* in the Leeds/Liverpool canal, from which the massive individual seen by Mr. Richard Freeman of the CFZ in 2002, and chronicled in my book *Monster of the Mere* (CFZ 2002) have been ignored.

Other omissions include the beavers reported from Essex in the 1970s, the semi-wild population of black swans which has existed at Dawlish in South Devon for decades, and the colony of blue and gold macaws on Woodbury Common, again in South Devon.

He also misses several well-known occurrences of the common raccoon *(Procyon lotor)* including one killed in February 2009 in Hampshire, body parts found in Herne Bay, Kent in September 2008, and an incident some years ago in Exwick, Exeter where one was caught in a suburban garden. Whilst I agree with him that despite the lack of direct evidence, these animals are probably, in fact almost certainly, breeding in the United Kingdom, I would also humbly suggest that coatis *(Nasua* sp.) have also bred in the UK.

From a personal point of view, although his division of the animals in the book into:

> *Naturalised species*
> *Feral domestic species*
> *Reintroduced species*
> *Ephemeral species*

makes a lot of sense, I personally would have preferred to have seen the whole book laid out in phylogenetic order as was the original version. If it had been done like this, with – perhaps – a logo or page tab to delineate the different categories, it would – I think – have made for a more convenient reference book.

I would have liked this book to have taken much more time in presenting all the information available in a more easily approachable manner. For some reason, and I am finding it hard to analyse why, the original edition is still a more satisfying holistic read. But, it has been out of print for many years, this new edition

contains much information that I had not read before and whilst I would have liked to have a two volume set collecting together the information that it does and that which although published in mainstream newspapers and even government reports, seems to be more the domain of the Fortean zoological community. But like an Irishman called Paddy who apparently hailed from the town of Donegal and featured in a rude song my ex-sailor father taught me at around about the same time as I read Lever's original volume, it may not be as good as two but it's better than none at all. JD

Ed.s note: When I first found that there was a new edition (well, technically it's not a new edition, and it is never really referred to as such within the book) of this landmark book I got so excited I had to sit in a darkened room for ten minutes before ringing Jon to let him know. We both bought a copy, and anxiously waited…

The post being a little strange in Bristol, I received mine at nearly six in the evening. I put the kettle on for a coffee, and sat down to tear through the book, eager to see how distribution patterns of various species had changed in 35 years and, more importantly for me, to see what new species had been added to Lever's list.

I was sorely disappointed. Jon has covered a lot of the things that irritated me, but something that really irritated me, because it is a personal bug-bear, was the random, unexplained changes in the keys to the distribution maps. Why should mammals have the typical dark "blobs" which show distribution, but herps and birds have small dots scattered around showing each location, which ends up looking ridiculous when the distribution is so massive, such as in pheasants. Fish get it even worse. The colonies are not even marked; a small dot is instead present on the county where they were recorded which is not of much use when you need the exact placement of a colony. Another real irritation is the lack of photographs of a couple of species. Missing a photograph of a species that is present in a couple of tiny colonies and seems not to be doing much harm is one thing, but not having a photograph of a potentially highly invasive fish which people *need* to be able to identify is inexcusable.

I could probably have filled this entire issue of TAN ranting about the book, but the fact remains that when he (Lever) tells you something totally new and of real interest, it still has the power to send me sprinting into my room to research it in a fury of typing and web-surfing. The printing is of high quality (if you like glossy paper), the photographs, where present, are of superb quality, and the text is easily readable both at all levels of interest and knowledge. A mixed review then. My recommendation then would be to buy the first edition, and then really think hard if you need this new one. I am sure that in the fullness of time, both Jon and myself will find ourselves using it to fill the 35 year gap, but the fact still remains, this book is a missed opportunity. MB

*The History of British Birds*
D. W. Yalden & U. Albarella
**Paperback:** 272 pages
**Publisher:** OUP Oxford (5 Nov 2009)
**Language** English
**ISBN-10:** 0199581169
**ISBN-13:** 978-0199581160

I, for one, have always been interested in looking at biology from a long termed perspective, something that is totally against my human nature. Humans have short life spans (not for animals, but in geological and even archaeological time) and this makes it really hard to appreciate that the world is in a constant state of flux. Climates change and shift and habitats change latitudes slowly, but steadily. These shifting climates naturally force animals to move along with them to follow the climates which they are adapted to. As such, because of the shifting climate, new birds are beginning to breed in the British Isles as they more North as the climate warms. Little bittern and various egrets are breeding here now, and other birds (like little bustards) look set to soon move into the British breeding lists.

Because it looks at birds in "deep time" (geological and archaeological time), "The History of British Birds" is invaluable for anyone wishing to gain a better insight into the ever changing state of our British birds. Our current avifauna is considerably reduced from what it was even 500 years ago in terms of the larger species, but the total number of breeding birds is as high as it has ever been, if not higher. *The History...* contains a great introductory section on bird bone identification, before it ploughs through geological time up to the present day.

The text is scientific and accurate, but not unreadable to an interested member of the general public with a basic grasp of biology. Figures and diagrams are used throughout to illustrate key points, and these are well explained and referenced within the text.

Though at first glance it may seem as if the authors are paying more attention to the larger, more spectacular birds (like sea eagles, cranes and great auks), it must be born in mind that large birds have large bones, and the larger a bone is, the more likely it is to be preserved throughout time. Bird bones are light and are much more susceptible to being destroyed than mammal bones, so I cannot find any fault with looking at larger species.

An appendix at the end shows an annotated historical list of British bird specimens followed by a large reference section, giving anyone the opportunity to do their own research on a topic.

I wish the hardback edition was actually affordable, but at £60 I can only see it being bought by universities and large libraries. The paperback is less than half this price, and is certainly worth it. The facts and figures are great for whipping out in conversation with other natural historians in the pub, or somewhere similar, but it is the readability of the text, combined with the fascinating story of our avifauna that kept me enthralled throughout. MB

MADEIRA'S NATURAL HISTORY IN A NUTSHELL by Peter Sziemer

**Hardcover**
**Publisher:** Francisco Ribeiro, (2000)
**ISBN-10:** 9729177317
**ISBN-13:** 978-9729177316

When I was a child one of the things that my parents found very funny in the 1960s and 70s was a book called *English as She is Spoke* allegedly by José da Fonseca and Pedro Carolino and was – at least according to Stephen Pile in *The Book of Heroic Failures* in 1979 - a *bona fide* Portuguese to English phrasebook written in the 19th Century by two folk neither of whom could speak English, and who had used a French/English dictionary to translate an earlier (and apparently completely competent) Portuguese phrasebook written by da Fonseca alone. It was republished and sold to people like my parents in the middle years of the 20th Century in order to prove that Johnny Foreigner was indeed a rum cove.

We now live in far more enlightened times. I, for example, with the aid of those jolly nice fellows at Babel Fish have written enough drivel in bad Spanish on the subject of the chupacabras to be perfectly aware of the old adage that those who live in grass-houses shouldn't get stoned (or something like that) but it would be amiss of me not to note that the English usage in this often charming little book tends towards the eccentric, and the Union Jack on the top right hand corner suggests that the book has been translated into a number of different languages, presumably so the book can be sold to tourists of differing nationalities.

But on to the book. In recent months I have become more than mildly obsessed with the natural history of what has fairly recently become known as Macaronesia – the five archipelagos in the western north Atlantic, which are owned by three countries: Portugal, Spain and the autonomous Cape Verde Islands.

Four of the five archipelagos are inhabited, but interestingly apart from the Canary Islands - which were settled by Spain in 1402 and have been Spanish ever since – the others owned (or in the case of Cape Verde formerly owned) by Portugal had no indigenous peoples and therefore – as far as I am aware – are the only colonies in the history of European expansionism with no literal or figurative blood on their hands.

All of these islands have interesting, albeit sparse, biomes, and in many cases are home to interesting endemic species. Madeira, which was first discovered in 1418, has a number of endemic butterflies, three endemic birds, an endemic lizard, only one species of amphibian (and that introduced) and only one native freshwater fish – the common eel although several species of salmonid have been introduced. Its mammalian fauna is mostly notable for one endemic bat species and a thriving colony of one of the rarest seals in the world – the Mediterranean monk seal (*Monachus monachus*). However, the most important natural resource, at least in terms of biodiversity, is its population of terrestrial molluscs. Whereas in the United Kingdom, which is approximately 400 times the size of Madeira, there are only 116 snail species, the Madeiran archipelago is home to more than 270 species, covering 65 genera and 20 families, and new species are being discovered all the time. Over 70% of these endemics.

One of the most interesting are the huge glass snails in the sub-genus *Madeirovitrina*.

The author, Peter Szeimer, is a freelance biologist who hails from Vienna. In this lovely little book he introduces us to all the main species of animal and plant from the islands and recommends places to go, walks to be walked, and locations that the avid visitor really shouldn't miss.

I think it is a fair bet that my current obsession with Macaronesia will lead us to be visiting Madeira sooner rather than later, and this book will definitely be part of my hand luggage. JD

# BRISTOL

bluereef
AQUARIUM

# AQUARIUM REVIEW

The Blue Reef aquarium chain currently operates five aquariums in the UK, having recently (October 2009) opened its latest aquarium in Bristol. The building, in the heart of the @Bristol complex, takes over from the commercially unsuccessful WildWalk and makes use of the old building, adapting it to suit the aquarium, but retaining part of the walk itself to give the public a surface view of two of the larger tanks in the aquarium. I met up with Michael Coe, the Marketing Executive, and Daniel De Castro, the Zoological Manager, for a quick guided tour of the aquarium.

Arranged into three main areas: British marine species, tropical marines, and freshwater species from around the world, the aquarium has a large variety of species from radically different biomes; from rock pools, brackish mangroves and the Amazon, to sunken wrecks, coral reefs and African lakes. A nice touch is that the largest tank in the aquarium is not for tropical species, but is for a large range of native marines. The aquarium manages a superb balance between showing off charismatic species that interest the public, whilst also reserving places for more obscure species to interest "those-in-the-know". It's not often that I go to an aquarium and am greeted by species that not only have I never seen before, but had no idea even existed; the fact that the false Anableps (*Rhinomugil corsula*) had me rushing to the literature once home to look it up is a very dear complement!

Regarding the largest coral reef tank, Daniel said: "The decision to choose smaller species of shark for

A mouthbrooding yellow Malawi cichlid; *(Labidochromis caeruleus)* if you look carefully you can see the eye of the baby hidden inside the parent's mouth.

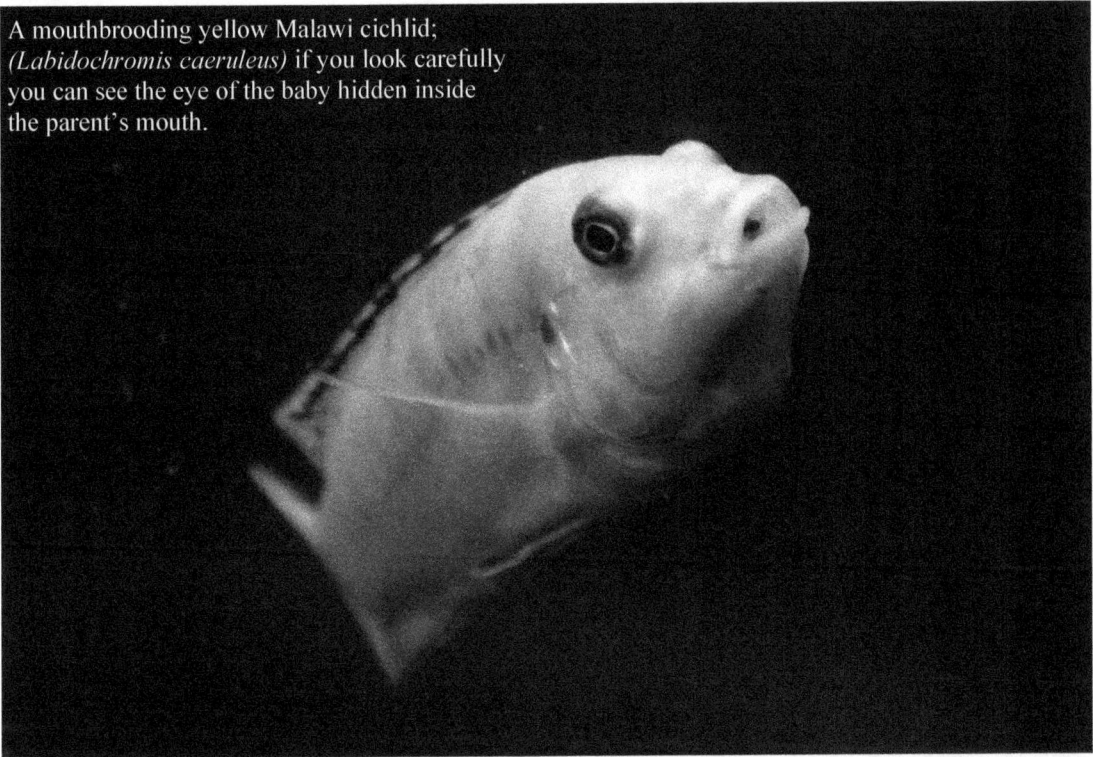

the aquarium reduced the minimum size of the fish that could also inhabit the tank, as well as allowing more rockwork to be put in. This allowed us to produce a tank that more accurately showed a coral reef with a great expanse of rockwork, but also with a wide variety of fish species showing a huge range of sizes, shapes and colours." He's right, it works! It really was refreshing to visit an aquarium that did not view sand tiger sharks as the be all and end all of sharks in aquaria. Though the Amazon river tank is pretty much a tank for tankbusters (large species of freshwater fish that people dump off to aquariums when they get too big for their own tanks) the raised walkway surrounding it provides an extra dimension into what could have been a very generic tank. As easily as I could rant about naïve people buying cute baby fish which rapidly grow into three foot monsters, I do still love seeing red-tailed catfish in huge tanks where they can stretch out. Other tanks and fish which deserve mention include the brackish tank which manages to avoid the usual monos, scats and archers in favour of smaller species like orange chromides (one of my all time favourite cichlids), mudskippers and the aforementioned false Anableps; the tall tank containing ballan wrasse (opposite) and conger eels had to be the tank I most wanted to take home, as it were; and the baby horseshoe crabs in one of the tanks were certainly the most enjoyable animals to watch.

Conservation and breeding are important aspects of running a successful aquarium, and indeed, the aquarium is currently breeding seahorses, *Geophagus* cichlids, dogfish and pipefish, but more species will begin breeding as they settle down (the large colony of threatened Banggai cardinalfish will surely start breeding). So, on all levels, the Blue Reef aquarium at Bristol is really worth a visit. A wide range of fish and environments are displayed in clean, well kept tanks (the filtration units are stunning), and with a strong attitude towards animal welfare and conservation, Bristol's Blue Reef aquarium deserves your attention. **Max Blake.**

# THE AQUARIUM GAZETTE

The Aquarium Gazette
16 Potter Hill, Pickering, N.Yorks.  Y018 8AA
Telephone 01751 472715
Email aquariumgazette@yahoo.com
Website www.theaquariumgazette.co.uk

Edited by The Amateur Naturalist aquarium fish contributor David Marshall, The Aquarium Gazette is the U.K.'s first bi-monthly aquarium magazine to be produced for computer users.  The Aquarium Gazette is available in both Word 2000 and PDF format, which can be purchased either on disc or for sending directly (PDF only) to your computer inbox.  The PDF format can be transferred to the e-books system without any problem.

Each Issue contains a wealth of aquatic knowledge and is comprised of between 10 and 13 varied feature articles, written by well known Aquarists', plus news and views from Aquatic Societies and Zoological Gardens etc.

Alongside articles on fish we have featured a number of articles about 'other creatures' that inhabit 'watery worlds', excellent articles on planted aquaria, are dedicated to aquatic conservation and have delved into the mysteries of aquatic cryptozoology.

Photograph copyright of J. Goulder

One of our much acclaimed non-fish articles was a world exclusive on the breeding of Poison Dart Frogs that features photographs of the youngsters at various stages of growth.

The cost, on disc, of a single Issue is £3.80, inclusive of postage and packing, while the e-mail version can be directly transferred to your e-mail inbox at a cost of £3. Details of obtaining compilation discs can be found on our website.

Payment can be made by sending a cheque (made payable to The Aquarium Gazette) to the above address or through our Paypal account (aquariumgazette@yahoo.com).  If your preferred method of payment is through Paypal then please add 25p for a single Issue of the disc format.

# THE CENTRE FOR FORTEAN ZOOLOGY

## So, what is the Centre for Fortean Zoology?

We are a non profit-making organisation founded in 1992 with the aim of being a clearing house for information, and coordinating research into mystery animals around the world. We also study out of place animals, rare and aberrant animal behaviour, and Zooform Phenomena; little-understood "things" that appear to be animals, but which are in fact nothing of the sort, and not even alive (at least in the way we understand the term).

## Why should I join the Centre for Fortean Zoology?

Not only are we the biggest organisation of our type in the world, but - or so we like to think - we are the best. We are certainly the only truly global Cryptozoological research organisation, and we carry out our investigations using a strictly scientific set of guidelines. We are expanding all the time and looking to recruit new members to help us in our research into mysterious animals and strange creatures across the globe. Why should you join us? Because, if you are genuinely interested in trying to solve the last great mysteries of Mother Nature, there is nobody better than us with whom to do it.

## What do I get if I join the Centre for Fortean Zoology?

For £12 a year, you get a four-issue subscription to our journal *Animals & Men*. Each issue contains 60 pages packed with news, articles, letters, research papers, field reports, and even a gossip column! The magazine is A5 in format with a full colour cover. You also have access to one of the world's largest collections of resource material dealing with cryptozoology and allied disciplines, and people from the CFZ membership regularly take part in fieldwork and expeditions around the world.

## How is the Centre for Fortean Zoology organised?

The CFZ is managed by a three-man board of trustees, with a non-profit making trust registered with HM Government Stamp Office. The board of trustees is supported by a Permanent Directorate of full and part-time staff, and advised by a Consultancy Board of specialists - many of whom are world-renowned experts in their particular field. We have regional representatives across the UK, the USA, and many other parts of the world, and are affiliated with other organisations whose aims and protocols mirror our own.

## I am new to the subject, and although I am interested I have little practical knowledge. I don't want to feel out of my depth. What should I do?

Don't worry. We were *all* beginners once. You'll find that the people at the CFZ are friendly and approachable. We have a thriving forum on the website which is the hub of an ever-growing electronic community. You will soon find your feet. Many members of the CFZ Permanent Directorate started off as ordinary members, and now work full-time chasing monsters around the world.

## I have an idea for a project which isn't on your website. What do I do?

Write to us, e-mail us, or telephone us. The list of future projects on the website is not ex-haustive. If you have a good idea for an investigation, please tell us. We may well be able to help.

## How do I go on an expedition?

We are always looking for volunteers to join us. If you see a project that interests you, do not hesi-tate to get in touch with us. Under certain circumstances we can help provide funding for your trip. If you look on the future projects section of the website, you can see some of the projects that we have pencilled in for the next few years.

In 2003 and 2004 we sent three-man expeditions to Sumatra looking for Orang-Pendek - a semi-legendary bipedal ape. The same three went to Mongolia in 2005. All three members started off merely subscribers to the CFZ magazine.

Next time it could be you!

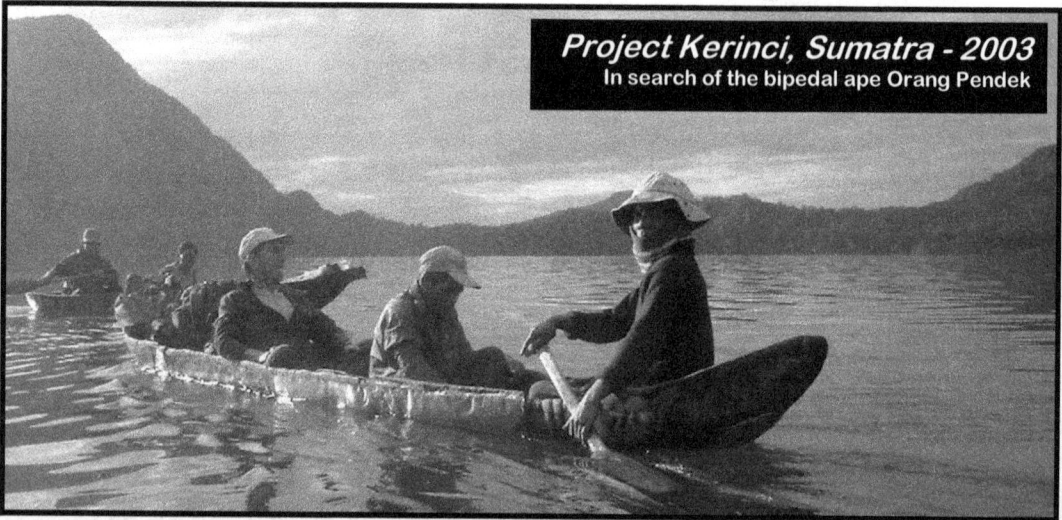

### Project Kerinci, Sumatra - 2003
In search of the bipedal ape Orang Pendek

## How is the Centre for Fortean Zoology funded?

We have no magic sources of income. All our funds come from donations, membership fees, works that we do for TV, radio or magazines, and sales of our publications and merchandise. We are al-ways looking for corporate sponsorship, and other sources of revenue. If you have any ideas for fund-raising please let us know. However, unlike other cryptozoological organisations in the past, we do not live in an intellectual ivory tower. We are not afraid to get our hands dirty, and further-more we are not one of those organisations where the membership have to raise money so that a privileged few can go on expensive foreign trips. Our research teams, both in the UK and abroad, consist of a mixture of experienced and inexperienced personnel. We are truly a community, and work on the premise that the benefits of CFZ membership are open to all.

## What do you do with the data you gather from your investigations and ex-peditions?

Reports of our investigations are published on our website as soon as they are available. Prelimi-nary reports are posted within days of the project finishing.

Each year we publish a 200 page yearbook containing research papers and expedition reports too long to be printed in the journal. We freely circulate our information to anybody who asks for it.

No. Each year since 2000 we have held our annual convention - the *Weird Weekend* - in Exeter. It is three days of lectures, workshops, and excursions. But most importantly it is a chance for members of the CFZ to meet each other, and to talk with the members of the permanent directorate in a relaxed and informal setting and preferably with a pint of beer in one hand. Since 2006 - the *Weird Weekend* has been bigger and better and held on the third weekend in August in the idyllic rural location of Woolsery in North Devon.

Since relocating to North Devon in 2005 we have become ever more closely involved with other community organisations, and we hope that this trend will continue. We also work closely with Police Forces across the UK as consultants for animal mutilation cases, and we intend to forge closer links with the coastguard and other community services. We want to work closely with those who regularly travel into the Bristol Channel, so that if the recent trend of exotic animal visitors to our coastal waters continues, we can be out there as soon as possible.

We are building a Visitor's Centre in rural North Devon. This will not be open to the general public, but will provide a museum, a library and an educational resource for our members (currently over 400) across the globe. We are also planning a youth organisation which will involve children and young people in our activities.

Apart from having been the only Fortean Zoological organisation in the world to have consistently published material on all aspects of the subject for over a decade, we have achieved the following concrete results:

- Disproved the myth relating to the headless so-called sea-serpent carcass of Durgan beach in Cornwall 1975
- Disproved the story of the 1988 puma skull of Lustleigh Cleave
- Carried out the only in-depth research ever into the mythos of the Cornish Owlman
- Made the first records of a tropical species of lamprey
- Made the first records of a luminous cave gnat larva in Thailand
- Discovered a possible new species of British mammal - the beech marten
- In 1994-6 carried out the first archival fortean zoological survey of Hong Kong
- In the year 2000, CFZ theories were confirmed when an new species of lizard was added to the British list
- Identified the monster of Martin Mere in Lancashire as a giant wels catfish
- Expanded the known range of Armitage's skink in the Gambia by 80%
- Obtained photographic evidence of the remains of Europe's largest known pike
- Carried out the first ever in-depth study of the *ninki-nanka*
- Carried out the first attempt to breed Puerto Rican cave snails in captivity
- Were the first European explorers to visit the `lost valley` in Sumatra
- Published the first ever evidence for a new tribe of pygmies in Guyana
- Published the first evidence for a new species of caiman in Guyana
- Filmed unknown creatures on a monster-haunted lake in Ireland for the first time
- Had a sighting of orang pendek in Sumatra in 2009
- Published some of the best evidence ever for the almasty in southern Russia
- In the year 2010, CFZ theories were confirmed when relict populations of pine martens were found in various parts of southern and central England

## EXPEDITIONS & INVESTIGATIONS TO DATE INCLUDE:

- 1998 Puerto Rico, Florida, Mexico *(Chupacabras)*
- 1999 Nevada *(Bigfoot)*
- 2000 Thailand *(Giant snakes called nagas)*
- 2002 Martin Mere *(Giant catfish)*
- 2002 Cleveland *(Wallaby mutilation)*
- 2003 Bolam Lake *(BHM Reports)*
- 2003 Sumatra *(Orang Pendek)*
- 2003 Texas *(Bigfoot; giant snapping turtles)*
- 2004 Sumatra *(Orang Pendek; cigau, a sabre-toothed cat)*
- 2004 Illinois *(Black panthers; cicada swarm)*
- 2004 Texas *(Mystery blue dog)*
- Loch Morar *(Monster)*
- 2004 Puerto Rico *(Chupacabras; carnivorous cave snails)*
- 2005 Belize *(Affiliate expedition for hairy dwarfs)*
- 2005 Loch Ness *(Monster)*
- 2005 Mongolia *(Allghoi Khorkhoi aka Mongolian death worm)*
- 2006 Gambia *(Gambo - Gambian sea monster , Ninki Nanka and  Armitage's skink*
- 2006 Llangorse Lake *(Giant pike, giant eels)*
- 2006 Windermere *(Giant eels)*
- 2007  Coniston Water *(Giant eels)*
- 2007 Guyana  *(Giant anaconda,  didi, water tiger)*
- 2008 Russia *(Almasty)*
- 2009 Sumatra *(Orang pendek)*
- 2009 Republic of Ireland *(Lake Monster)*
- 2010 Texas *(Blue Dogs)*

# THE WORLD'S WEIRDEST PUBLISHING COMPANY

ANIMALS & MEN
ISSUES 16-20
THE JOURNAL OF THE CENTRE FOR FORTEAN ZOOLOGY
NEW HORIZONS
Edited by Jon Downes

BIG CATS LOOSE IN BRITAIN

PREDATOR DEATHMATCH
NICK MOLLOY
WITH ILLUSTRATIONS BY ANTHONY WALLIS

...TER!
...E ZOO... M PHENOMENA

Edited by Jonathan Downes and Richard Freeman

FOREWORD BY Dr. KARL SHUKER

A DAINTREE DIARY
Tales from Travels ... Daintree
tropical North ...nsland A...
CARL PORTMAN

STAR STEEDS

THE COLLECTED POEMS
Dr Karl P. N. Shuker

STRANGELYSTRANGE
...ly normal
an anthology of writings by
ANDY ROBERTS

# HOW TO START A PUBLISHING EMPIRE

Unlike most mainstream publishers, we have a non-commercial remit, and our mission statement claims that "we publish books because they deserve to be published, not because we think that we can make money out of them". Our motto is the Latin Tag "Pro bona causa facimus" (we do it for good reason), a slogan taken from a children's book `The Case of the Silver Egg` by the late Desmond Skirrow.

WIKIPEDIA: "The first book published was in 1988. `Take this Brother may it Serve you Well` was a guide to Beatles bootlegs by Jonathan Downes. It sold quite well, but was hampered by very poor production values, being photocopied, and held together by a plastic clip binder. In 1988 A5 clip binders were hard to get hold of, so the publishers took A4 binders and cut them in half with a hacksaw. It now reaches surprisingly high prices second hand.

The production quality improved slightly over the years, and after 1999 all the books produced were ringbound with laminated colour covers. In 2004, however, they signed an agreement with LightningSource, and all books are now produced perfect bound, with full colour covers."

Until 2010 all our books, the majority of which are/were on the subject of mystery animals and allied disciplines, were published by `CFZ Press`, the publishing arm of the Centre for Fortean Zoology (CFZ), and we urged our readers and followers to draw a discreet veil over the books that we published that were completely off topic to the CFZ.

However, in 2010 we decided that enough was enough and launched a second imprint, `Fortean Words` which aims to cover a wide range of non animal-related esoteric subjects. Other imprints will be launched as and when we feel like it, however the basic ethos of the company remains the same: Our job is to publish books and magazines that we feel are worth publishing, whether or not they are going to sell. Money is, after all - as my dear old Mama once told me - a rather vulgar subject, and she would be rolling in her grave if she thought that her eldest son was somehow in `trade`.

Luckily, so far our tastes have turned out not to be that rarified after all, and we have sold far more books than anyone ever thought that we would, so there is a moral in there somewhere…

Jon Downes,
Woolsery, North Devon
July 2010

# CFZ PRESS

## Other Books in Print

*The Mystery Animals of Ireland* by Gary Cunningham and Ronan Coghlan
*Monsters of Texas* by Gerhard, Ken
*The Great Yokai Encyclopaedia* by Freeman, Richard
*NEW HORIZONS: Animals & Men issues 16-20 Collected Editions Vol. 4* by Downes, Jonathan
*A Daintree Diary -*
*Tales from Travels to the Daintree Rainforest in tropical north Queensland, Australia* by Portman, Carl
*Strangely Strange but Oddly Normal* by Roberts, Andy
*Centre for Fortean Zoology Yearbook 2010* by Downes, Jonathan
*Predator Deathmatch* by Molloy, Nick
*Star Steeds and other Dreams* by Shuker, Karl
*CHINA: A Yellow Peril?* by Muirhead, Richard
*Mystery Animals of the British Isles: The Western Isles* by Vaudrey, Glen
*Giant Snakes - Unravelling the coils of mystery* by Newton, Michael
*Mystery Animals of the British Isles: Kent* by Arnold, Neil
*Centre for Fortean Zoology Yearbook 2009* by Downes, Jonathan
*CFZ EXPEDITION REPORT: Russia 2008* by Richard Freeman *et al*, Shuker, Karl (fwd)
*Dinosaurs and other Prehistoric Animals on Stamps - A Worldwide catalogue* by Shuker, Karl P. N
*Dr Shuker's Casebook* by Shuker, Karl P.N
*The Island of Paradise - chupacabra UFO crash retrievals,*
*and accelerated evolution on the island of Puerto Rico* by Downes, Jonathan
*The Mystery Animals of the British Isles: Northumberland and Tyneside* by Hallowell, Michael J
*Centre for Fortean Zoology Yearbook 1997* by Downes, Jonathan (Ed)
*Centre for Fortean Zoology Yearbook 2002* by Downes, Jonathan (Ed)
*Centre for Fortean Zoology Yearbook 2000/1* by Downes, Jonathan (Ed)
*Centre for Fortean Zoology Yearbook 1998* by Downes, Jonathan (Ed)
*Centre for Fortean Zoology Yearbook 2003* by Downes, Jonathan (Ed)
*In the wake of Bernard Heuvelmans* by Woodley, Michael A
*CFZ EXPEDITION REPORT: Guyana 2007* by Richard Freeman *et al*, Shuker, Karl (fwd)
*Centre for Fortean Zoology Yearbook 1999* by Downes, Jonathan (Ed)
*Big Cats in Britain Yearbook 2008* by Fraser, Mark (Ed)
*Centre for Fortean Zoology Yearbook 1996* by Downes, Jonathan (Ed)
*THE CALL OF THE WILD - Animals & Men issues 11-15*
*Collected Editions Vol. 3* by Downes, Jonathan (ed)

*Ethna's Journal* by Downes, C N
*Centre for Fortean Zoology Yearbook 2008* by Downes, J (Ed)
*DARK DORSET -Calendar Custome* by Newland, Robert J
*Extraordinary Animals Revisited* by Shuker, Karl
*MAN-MONKEY - In Search of the British Bigfoot* by Redfern, Nick
*Dark Dorset Tales of Mystery, Wonder and Terror* by Newland, Robert J and Mark North
*Big Cats Loose in Britain* by Matthews, Marcus
*MONSTER! - The A-Z of Zooform Phenomena* by Arnold, Neil
*The Centre for Fortean Zoology 2004 Yearbook* by Downes, Jonathan (Ed)
*The Centre for Fortean Zoology 2007 Yearbook* by Downes, Jonathan (Ed)
*CAT FLAPS! Northern Mystery Cats* by Roberts, Andy
*Big Cats in Britain Yearbook 2007* by Fraser, Mark (Ed)
*BIG BIRD! - Modern sightings of Flying Monsters* by Gerhard, Ken
*THE NUMBER OF THE BEAST - Animals & Men issues 6-10*
*Collected Editions Vol. 1* by Downes, Jonathan (Ed)
*IN THE BEGINNING* - Animals & Men *issues 1-5 Collected Editions Vol. 1* by Downes, Jonathan
*STRENGTH THROUGH KOI - They saved Hitler's Koi and other stories* by Downes, Jonathan
*The Smaller Mystery Carnivores of the Westcountry* by Downes, Jonathan
*CFZ EXPEDITION REPORT: Gambia 2006* by Richard Freeman *et al*, Shuker, Karl (fwd)
*The Owlman and Others* by Jonathan Downes
*The Blackdown Mystery* by Downes, Jonathan
*Big Cats in Britain Yearbook 2006* by Fraser, Mark (Ed)
*Fragrant Harbours - Distant Rivers* by Downes, John T
*Only Fools and Goatsuckers* by Downes, Jonathan
*Monster of the Mere* by Jonathan Downes
*Dragons:More than a Myth* by Freeman, Richard Alan
*Granfer's Bible Stories* by Downes, John Tweddell
*Monster Hunter* by Downes, Jonathan

# Fortean Words

The Centre for Fortean Zoology has for several years led the field in Fortean publishing. CFZ Press is the only publishing company specialising in books on monsters and mystery animals. CFZ Press has published more books on this subject than any other company in history and has attracted such well known authors as Andy Roberts, Nick Redfern, Michael Newton, Dr Karl Shuker, Neil Arnold, Dr Darren Naish Jon Downes, Ken Gerhard and Richard Freeman.

Now CFZ Press are launching a new imprint. Fortean Words is a new line of books dealing with Fortean subjects other than cryptozoology, which is - after all - the subject the CFZ are best known for. Fortean Words is being launched with a spectacular multi-volume series called *Haunted Skies* which covers British UFO sightings between 1940 and 2010. Former policeman John Hanson and his long-suffering partner Dawn Holloway have compiled a peerless library of sighting reports, many that have been made public before.

Other forthcoming books include a look at the Berwyn Mountains UFO case by renowned Fortean Andy Roberts and a series of books by transatlantic research Nick Redfern.

CFZ Press are dedicated to maintaining the fine quality of their works with Fortean Words. New authors tackling new subjects will always be encouraged, and we hope that our books will continue to be as ground breaking and popular as ever.

www.ingramcontent.com/pod-product-compliance
Lightning Source LLC
Chambersburg PA
CBHW051415200326
41520CB00023B/7243